Green Manufacturing and Materials Processing Methods

In this modern technological era, conserving and making better use of resources like energy, water, and other essential resources have recently been one of the main concerns for the manufacturing industry. To successfully compete against the competition, industries are replacing outdated manufacturing techniques with cutting-edge ones that are sustainable in terms of cost, energy usage, better product quality, and environmental safety. Green manufacturing has become one of the key priorities for attaining this.

Green Manufacturing and Materials Processing Methods: Characterizations, Applications, and Design offers a critical review of the past work done in green manufacturing and material processing technologies. It presents recent research and development that is going on currently with green manufacturing techniques and discusses characterizations, applications, and the design aspect of materials processed through green manufacturing technologies. With a focus on the sustainability aspect, this book showcases new breakthroughs and comparisons of cutting-edge sustainable manufacturing and materials processing with currently available conventional methods. Highlights throughout the book are on improvements used in various manufacturing processes such as casting, joining, drilling, surface engineering, sintering, and composite manufacturing.

This book will serve as a first-hand information source for academic researchers and industrial firms. With the help of this book, readers will have a unique opportunity to comprehend and evaluate recent advancements in green manufacturing and material processing technology. This book will be the go-to resource for individuals who desire to do research or development in the area of sustainable manufacturing and material processing technologies.

Sustainable Manufacturing Technologies: Additive, Subtractive, and Hybrid

Series Editors: Chander Prakash, Sunpreet Singh, Seeram Ramakrishna, and Linda Yongling Wu

This book series offers the reader comprehensive insights of recent research breakthroughs in additive, subtractive, and hybrid technologies while emphasizing their sustainability aspects. Sustainability has become an integral part of all manufacturing enterprises to provide various techno-social pathways toward developing environmental friendly manufacturing practices. It has also been found that numerous manufacturing firms are still reluctant to upgrade their conventional practices to sophisticated sustainable approaches. Therefore this new book series is aimed to provide a globalized platform to share innovative manufacturing mythologies and technologies. The books will encourage the eminent issues of the conventional and non-conventual manufacturing technologies and cover recent innovations.

Handbook of Post-Processing in Additive Manufacturing
Requirements, Theories, and Methods
Edited by Gurminder Singh, Ranvijay Kumar, Kamalpreet Sandhu, Eujin Pei, and Sunpreet Singh

Manufacturing Engineering and Materials Science
Tools and Applications
Edited by Abhineet Saini, B. S. Pabla, Chander Prakash, Gurmohan Singh, Alokesh Pramanik

Manufacturing Technologies and Production Systems
Principles and Practices
Edited by Abhineet Saini, B. S. Pabla, Chander Prakash, Gurmohan Singh, Alokesh Pramanik

Modern Hybrid Machining and Super Finishing Processes
Technology and Applications
Edited by Ankit Sharma, Amrinder S. Uppal, Bhargav P. Prajwal, Atul Babbar and Chander Prakesh

Green Manufacturing and Materials Processing Methods
Characterizations, Applications, and Design
Edited by Sarbjeet Kaushal, Sandeep Bansal, Chander Prakash, Bhupinder Singh, and Dheeraj Gupta

For more information on this series, please visit: www.routledge.com/Sustainable-Manufacturing-Technologies-Additive-Subtractive-and-Hybrid/book-series/CRCSMTASH

Green Manufacturing and Materials Processing Methods

Characterizations, Applications, and Design

Edited by
Sarbjeet Kaushal, Sandeep Bansal,
Chander Prakash, Bhupinder Singh, and
Dheeraj Gupta

CRC Press is an imprint of the
Taylor & Francis Group, an informa business

First edition published 2025
by CRC Press
2385 NW Executive Center Drive, Suite 320, Boca Raton FL 33431

and by CRC Press
4 Park Square, Milton Park, Abingdon, Oxon, OX14 4RN

CRC Press is an imprint of Taylor & Francis Group, LLC

ISBN: 978-1-032-58017-3 (hbk)
ISBN: 978-1-032-58241-2 (pbk)
ISBN: 978-1-003-44922-5 (ebk)

DOI: 10.1201/9781003449225

Typeset in Times
by Newgen Publishing UK

Contents

Preface

The incorporation of green manufacturing practices is crucial for maintaining the long-term economic competitiveness and technical advancement of the current society and the next generation. This book explores innovative engineering methods for production and offers a transition from conventional to environmentally friendly manufacturing. The numerous green manufacturing and material processing methods for advanced materials are covered in this book. This book is going to be a thorough and reliable source on producing metals, ceramics, and their composites in a sustainable manner. In their respective fields, the authors are all acknowledged as active scholars and practitioners. We anticipate that this book will offer a roadmap for future advancement in the field of environmentally friendly manufacturing and material processing techniques. Additionally, we hope that this book will satisfy the academic and research needs of students throughout.

About the Editors

Sarbjeet Kaushal is currently working as a professor in the Department of Mechanical Engineering at Gulzar Group of Institutions, Khanna, Ludhiana, Punjab, India. He received his Ph.D. in Mechanical Engineering from Thapar Institute of Engineering and Technology, Patiala, India. His research interests include sustainable manufacturing, surface engineering, microwave processing of materials, tribology of composite materials, and advanced manufacturing processes. He has published more than 50 papers in referred journals including 26 SCI/SCIE indexed journals. He has organized several seminars, workshops, and conferences at national and international level. He has also published two books. He is the reviewer of several peer-reviewed journals such as *Vacuum* (Elsevier), *International Journal of Metal Castings* (Springer), and *Journal of Engineering Manufacture* (SAGE).

Sandeep Bansal is working as Assistant Professor in the Department of Mechanical Engineering at Shree Guru Gobind Singh Tricentenary University, Gurgaon, India. He obtained his Diploma in Mechanical Engineering from S.L.I.E.T, Longowal, Sangrur, India; B.Tech. in Mechanical Engineering from the Institution of Engineers India; Master's in Mechanical Engineering from Bhai Gurdas Institute of Engineering and Technology, Sangrur, India; and Doctoral Degree from Thapar Institute of Engineering and Technology, Patiala, India. He has more than 18 years of teaching and industrial experience. He has published more than 15 articles in journals and conferences. He worked on a DST (SERB)-funded research project during his Ph.D. His areas of interest are microwave material processing, surface engineering, tribology, and biomedical.

Chander Prakash is Pro-Vice Chancellor of Research & Development at Chandigarh University, Punjab, India. He has published over 400 (330 SCI/Scopus) scientific articles in peer-reviewed, reputable, top-notch journals, conferences, and books. Prakash is a highly cited researcher at the international level, and he has 8874 citations, an H-index of 53, and an i10 index of 182. He is among India's Top 1% of leading Mechanical and Aerospace Engineering scientists, Research.com. He holds 38 ranks in India. He has edited 23 books and 3 authored books for various reputed publishers like Springer, Elsevier, CRC Press, and World Scientific. He is a series editor of the book *Sustainable Manufacturing Technologies: Additive, Subtractive, and Hybrid*, CRC Press, Taylor & Francis, where more than 25 edited books were published by national and international researchers. He is serving as an editorial board member of peer-reviewed international journals *Cogent Engineering* and *Frontiers in Manufacturing Technology*. He is serving as a guest editor for peer-reviewed SCI-indexed journals.

Bhupinder Singh is currently working as a Research Fellow in the Department of Chemical and Environmental Engineering, University of Nottingham, UK. He obtained his Ph.D from IIT Mandi, India; and Master's degree from Thapar Institute of Engineering and Technology, India. His areas of interest include surface engineering, cavitation erosion, and advanced manufacturing. He possesses a track record of designing successful experimental programs, including numerous research papers in reputed journals and a patent.

Dheeraj Gupta is currently Professor and Head of the Central Workshop at the Thapar Institute of Engineering and Technology, Patiala, India. He obtained his B.E. in Mechanical Engineering from Shri Govindram Seksaria Institute of Technology and Science, Indore, India; and Master's and Doctoral degrees from Indian Institute of Technology, Roorkee, India. He has more than 10 years of teaching and research experience. He has worked on a European Union funded project in Lisbon, Portugal. He has published more than 70 articles in journals and conferences. He has filed two Indian patents on microwave processing of metals. Currently, he is a principal investigator for an Indian government-funded project. His areas of interest are microwave material processing, surface engineering, tribology, biomedical, and advanced manufacturing. He has supervised many Master's thesis and six Ph.D thesis.

Contributors

Amit Bansal
Department of Mechanical Engineering, IKGPTU Kapurthala
India

Sandeep Bansal
Mechanical Engineering Department, SGT University
Gurgaon, Haryana, India

Dharmpal Deepak
Department of Mechanical Engineering, Punjabi University
Patiala, Punjab, India

Sulakshna Dwivedi
Jagat Guru Nanak Dev Punjab State Open University
Patiala, India

Ritu Bala Garg
Department of Civil Engineering, Punjabi University
Patiala, India

Dheeraj Gupta
Department of Mechanical Engineering, Thapar Institute of Engineering and Technology
Patiala, Punjab, India

Sarbjeet Kaushal
Department of Mechanical Engineering, Gulzar Group of Institutions
Khanna, Punjab, India

Kuldeepak
Department of Mechanical Engineering, Ph.D. Research scholar IKGPTU
and
Rayat Bahra Institute of Engineering & Nano-Technology
India

Sanjeev Kumar
Mechanical Engineering Department, Swami Vivekanand Subharti University
Meerut, Uttar Pradesh, India

Rohit Lade
Chemical Engineering Department, Parul University
India

Jatinder Pal
Department of Mechanical Engineering, Thapar Institute of Engineering and Technology
and
Department of Mechanical & Production Engineering, Guru Nanak Dev Engineering College Ludhiana, Punjab, India

Pankaj Rana
Department of Mechanical Engineering, Punjabi University
Patiala, Punjab, India

Bhupinder Singh
Department of Chemical and Environmental Engineering, University of Nottingham
Nottingham, United Kingdom

Charanjit Singh
Department of Mechanical Engineering, Punjabi University
Patiala, Punjab, India

Davinder Singh
Department of Mechanical Engineering, Punjabi University
Patiala, Punjab, India

Gurpreet Singh
Department of Civil Engineering, Punjabi University
Patiala, India

Kanwarjeet Singh
Mechanical Engineering Department, Delhi Skill and Entrepreneurship University
New Delhi, India

Sukhjinder Singh
Department of Mechanical Engineering, Punjabi University
Patiala, Punjab, India

Tejinder Paul Singh
LMT School of Management, Thapar Institute of Engineering and Technology, Dera Bassi Campus
Punjab, India

Yudhveer Kumar Verma
PhD Research Scholar Delhi Skill and Entrepreneurship University
New Delhi, India

1 Introduction to Green Manufacturing Techniques

Kuldeepak, Sarbjeet Kaushal, and Amit Bansal

1.1 INTRODUCTION

One of the most challenging aspects of the environment is global warming. According to a recent scientific investigation, human actions carried out for survival are the primary cause of global warming. All types of industries, including the cement and manufacturing sectors, engage in this activity. Because of industrialization, developed nations release more carbon into the atmosphere, contributing to global warming (Dilip Maruthi and Rashmi, 2015). The term "green manufacturing," also referred to as "environmentally conscious manufacturing for the environment," refers to a contemporary manufacturing technique that fully considers resource consumption and environmental impact to minimize environmental harm and maximize resource utilization throughout the entire production cycle, including design, manufacture, packaging, transportation, and reuse. The firm can gain additional social and economic benefits by implementing and optimizing green manufacturing. The green manufacturing system encompasses the entire product life cycle. Taking into account the environmental impact and resource consumption of modern manufacturing systems, the objective is to minimize environmental impact while ensuring that resources are used as efficiently as possible throughout the entire product life cycle, from design through production, packaging, transportation, use, and disposal (Tao and Zhao, 2016). Sustainability is the core aim of green manufacturing. For the benefit of future generations, every manufacturing sector should conserve resources. Additionally, they must be aware of when their obligations expire and the permitted level of hazardous emissions into the environment. Green production enhances public perception, reduces wasteful spending, and fosters innovation. The practice of "green manufacturing" entails making investments in bettering the production process rather than control technology, switching to renewable sources in place of finite ones, encouraging employee recycling, and letting businesses choose whether to create or purchase the product (Dilip Maruthi and Rashmi, 2015).

Numerous firms are adjusting their operations to achieve sustainability goals as a result of the growing awareness of the environmental imperative. The adoption of the UN Sustainable Development Goals in many corporations' plans and yearly reports

DOI: 10.1201/9781003449225-1

serves as clear evidence of this (Paul et al., 2014). There is a steadfast consensus that the manufacturing industries are still working in a largely linear and unsustainable manner despite the expanding amount of scientific knowledge. To motivate industry to take environmental action, external incentives are still necessary. By 2030, greenhouse gas emissions must be reduced by 55% from 1990 levels, according to the European Climate Law (Despeisse et al., 2022).

In this chapter, the author tries to explore the various studies done in the domain of green manufacturing and techniques for the implementation of green manufacturing.

1.2 VARIOUS TECHNIQUES TO ACHIEVE GREEN MANUFACTURING

1.2.1 Lean Manufacturing

Numerous strategies are used in lean manufacturing. These methods aid in defining lean manufacturing's goals (Dilip Maruthi and Rashmi, 2015b). The strategies listed below are essential to lean manufacturing, as can be seen by taking a close look at it (Figure 1.1).

1.2.1.1 Just in Time

At the core of lean manufacturing lies the "pull" model, an approach based on demand-driven production. This methodology revolves around procuring supplies, manufacturing, and distributing goods only as necessary, minimizing excess inventory, and resource usage. Implementing this principle involves producing small, continuous batches of goods to maintain a smooth and efficient production flow. By reducing batch sizes, manufacturers can closely monitor quality and address any issues in real-time, significantly reducing the likelihood of subsequent batches having lower quality.

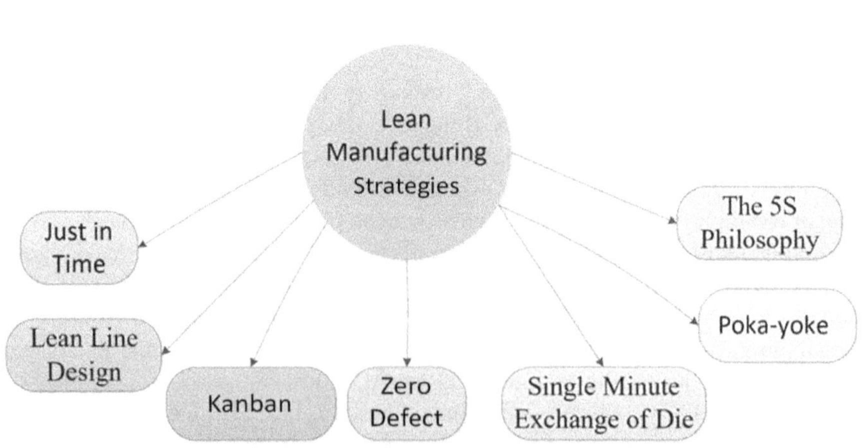

FIGURE 1.1 Lean manufacturing strategies.

1.2.1.2 Kanban

A key tool in manufacturing, kanban helps with this process by providing visual cues for when and what to produce, ensuring that resources are employed effectively, and keeping production in line with actual demand. Establishing regular signals that motivate activities like ordering, changing, or locating things in line with the Just In Time strategy is a crucial part of using people in lean manufacturing. By reducing excess inventory and guaranteeing that things are accessible exactly when needed—avoid overproduction or surplus stock—seek to optimize operations. Teams may effectively manage resources, maintain a lean inventory and production system where items are replaced or obtained just as needed, optimize throughput, and cut waste by setting up these unambiguous signals within the system.

1.2.1.3 Zero Defects

The Zero Defects approach focuses on delivering high-quality results during the original manufacturing phase, rather than dedicating additional money and time to correct poor goods afterward. By employing this technique, personnel is taught that no flaw is acceptable, building a culture that prioritizes correctness and precision from the start. This approach motivates employees to do tasks methodically and properly the first time, avoiding the need for rework or correction while fostering a dedication to continually producing high-quality goods.

1.2.1.4 Single-Minute Exchange of Die

In the automobile industry, switching production lines from one car model to another usually takes several days. However, introducing flexibility into manufacturing is critical for adaptation. The Single-Minute Exchange of Die (SMED) technique optimizes assembly processes and equipment for fast and efficient transitions between production settings. The term "die" refers to the tools used to shape or create materials for specified components. SMED attempts to reduce the time necessary for these changeovers by simplifying the process and enabling quick transitions between manufacturing runs. This strategy improves manufacturing agility, allowing businesses to adapt quickly to changing market needs by minimizing downtime and optimizing production efficiency while switching between product lines or models.

1.2.1.5 The 5S Philosophy

Standardization is a vital component of lean manufacturing, highlighting the need for consistency and ease of use in tools, processes, and work environments. Companies avoid the requirement for excessive storage space for spare components and decrease potential points of failure by aiming for simple, standardized procedures. To attain and preserve a high degree of standardization in the production environment, a systematic technique known as the 5S System—Sort, Set in Order, Shine, Standardize, Sustain—can be implemented. This method makes sure that workspaces are tidy, processes are streamlined, and resources are used effectively, all of which support a lean and extremely productive manufacturing process.

1.2.1.6 Poka-yoke

Poka-yoke, which means "fool-proofing," is a crucial lean manufacturing technique that teaches workers how to operate equipment faultlessly. Its main goal is to prevent or correct human error, which will eventually lead to the elimination of product problems. When putting poka-yoke into practice, careful consideration of potential problems and the creation of workable solutions are required. Working together with quality assurance teams, engineers, and operators is essential to this process since it involves ongoing audits and efforts for improvement. Poka-yoke is divided into two subcategories: prevention and detection. The former focuses on preventing errors altogether, while the latter attempts to correct faults when they arise. Poka-yoke incorporates lean concepts by lowering waste, increasing customer value, and decreasing defects, and by emphasizing preventative steps as more efficient. Poka-yoke assists in attaining the goals of lean manufacturing by promoting an efficient, economical, and customer-focused production process.

1.2.1.7 Lean Line Design

Lean Line Design is a manufacturing methodology that integrates several ideas, including process orientation, ideal quality, standardization, flexibility, waste reduction, transparent processes, and employee involvement. It is particularly used in assembly lines. By assuring uniform quality, promoting adaptability within the production system, and streamlining workflow, this technique seeks to expedite assembly line operations. Lean Line Design emphasizes the active participation and engagement of employees while standardizing and eliminating waste to create transparent and efficient operations. Lean Line Design aims to increase overall productivity, reduce errors, and foster a continuous improvement culture in manufacturing assembly lines by incorporating these components.

1.2.2 Energy Efficiency and Renewable Energy

Zero Emission: At Zero Emission, we design technologies and techniques to maximize resource productivity and produce nearly no waste. Concepts for zero emissions can help with the following:

- Development of more efficient technology
- New manufacturing methods
- Conservation and recycling of natural resources (energy waste).

This objective can be achieved through a variety of strategies, including industrial ecology, pollution avoidance, technical innovation, byproduct synergy, cleaner production, and industrial ecology. All of the methods described above focus on reducing environmental and human health risks while disposing of trash or converting it into useful resources. From an environmental standpoint, eliminating waste is the best way to address pollution issues that pose a threat to ecosystems at the global, national, and local levels.

1.2.3 ISO 14000 and ISO 14001

The industries use these standards on a voluntary basis. They also encourage modifications that might be made to the production process that are environmentally friendly. Environmental management systems (EMS) standards such as ISO 14000 provide a framework for businesses to demonstrate their commitment to environmental responsibility. It is a tool for planning and carrying out operations that address environmental business concerns. Environmental disasters can be avoided thanks to ISO 14000 standards, which show good environmental management techniques and are effective at life-cycle analysis, environmental performance evaluation, and environmental labeling (Miranda et al., 2021).

The most significant regulation in the ISO 14000 series is the ISO 14001 standard. For small- to large-scale companies, ISO 14001 lists the specifications for an EMS. An EMS is a systematic method of resolving environmental problems within a sector. The Plan-Check-Do-Review-Improve cycle serves as the foundation for the ISO 14001 standard.

1.2.4 Reducing Emission by Pollutants

Within the European Union, road transportation accounts for over 72% of greenhouse gas emissions in the transportation sector, with air contributing 13.3%, marine 12.8%, and rail transportation around 0.5% (Miranda et al., 2021). As global environmental concerns escalate and the threat to the environment intensifies, there's a growing recognition of the imperative to prioritize environmental considerations in planning processes. Progressive societies prioritize controlling and evaluating polluting factors, instituting regulations, and formulating policies to prevent ecological issues (Bhoi et al., 2019; Miranda et al., 2021). Potential solutions include the following:

1. Shifting from air transportation to alternative modes with lower emissions.
2. Promoting the use of electric planes for shorter distances.
3. Advocating for wider adoption of larger aircraft for popular domestic routes and fuel-efficient planes for long international flights.
4. Potentially substituting some air passenger transport and short flights with high-speed railway services.

1.3 SUSTAINABLE MATERIAL SELECTION

Several researchers have explored material selection in sustainable product design. Holloway (Holloway, 1998) investigated a specific approach to material selection in mechanical design, focusing on material selection charts by Ashby and demonstrating their adaptability to incorporate environmental considerations. Ermolaeva et al. (2004) demonstrated the utilization of a structural optimization system to choose the optimal foams for car floor panel use in the bottom structure. They not only optimized for reduced mass and material cost but also evaluated the environmental impact of candidate materials throughout the entire life cycle of the structure.

FIGURE 1.2 Three dimensions of sustainability.

Source: Zarandi et al. (2011).

In recent years, the proportion of renewable energy sources has steadily increased. Concurrently, Earth's material resources are progressively becoming scarcer. The choice of appropriate materials is intricately linked with both product design and production system design (Leong et al., 2019; Stoffels et al., 2017).

The escalating depletion of natural resources and substantial ecological damage, adversely impacting global economic growth, societal well-being, and advancements in human health, has made sustainability a critical strategic consideration (Lin and Hao, 2020). Economic viability pertains to the financial dimensions of a project, while social concerns encompass considerations regarding human health and welfare. Natural or ecological issues refer to the stewardship of the environment. Figure 1.2 illustrates the three facets of sustainability and the trade-offs between them. Sustainability involves the interplay of three pillars: environmental, economic, and societal factors (Haleem et al., 2023). Addressing sustainability has become paramount in modern society, necessitating fundamental changes in cultures, structures, and practices to steer socio-technical systems toward more sustainable patterns of production and consumption (Paul et al., 2014).

The framework applied for integrating product and production engineering into material selection serves as a foundational model designed to establish analysis and synthesis techniques for diverse challenges in integrated product and production development. When choosing a material, one should take into account not only the functional performance needed for the application but also the financial and environmental effects that arise throughout the product life cycle. As a result, choosing sustainable materials may be thought of as a multiobjective challenge, including finding the optimum combination of accessible material profiles and sustainable product design needs. A sustainable material selection approach should assess a group of candidate materials and determine the "best material domains" (Ljungberg, 2007) by combining the three variables (social, economic, and environmental) (refer to Figure 1.3).

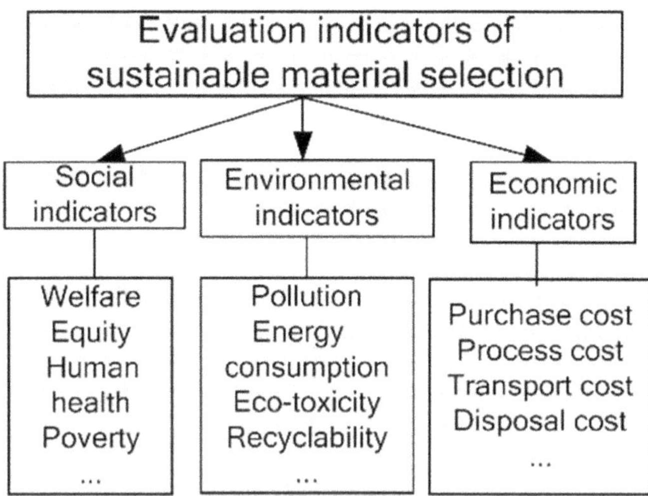

FIGURE 1.3 Evaluation indicators of sustainable material selection.

Source: Zarandi et al. (2011).

1.4 PRESENT SITUATION OF GREEN MANUFACTURING TECHNOLOGY

Due to the manufacturing sector's quick development, environmental concerns are receiving more and more attention from all nations. The "green wave" has forced the manufacturing sector to abandon its old methods and pursue green manufacturing.

Through a meticulous and timeless assessment, we scrutinized the utilization of green technology within the BRICS (Brazil, Russia, India, China, and South Africa) nations (Paul et al., 2014). These nations were chosen due to their considerable potential to impact the environment and their pivotal role in discussions regarding sustainable practices. Employing the Methodic Ordination technique, we curated a portfolio of relevant papers within this domain, totaling 170 studies. This approach enabled the identification of primary green technology practices employed across the BRICS countries. Noteworthy practices encompass sustainable agriculture, water management, waste handling, green energy initiatives, carbon reduction, eco-friendly construction, and policies regarding sustainability and eco-cities. China and India emerged with a larger volume of research due to their strides in sustainable development and heightened interest within their local academic communities. Despite potential, South Africa, Russia, and Brazil are yet to lead the global adoption of green technology practices.

Green manufacturing promotes environmental sustainability through the integration of green technology across a product's entire lifecycle, spanning research, production, utilization, and disposal. This approach involves the adoption of energy-efficient digital equipment such as servers, printers, projectors, and other devices. By minimizing waste during manufacturing, reducing power consumption in computer systems and peripherals, and implementing environmentally responsible disposal methods, green manufacturing significantly curtails the environmental impact of operations. Additionally, Green Computing presents promising prospects for environmental conservation. Automation and artificial intelligence are revolutionizing the recycling process, making it less labor-intensive and more efficient, allowing businesses to acquire, process, and dismantle complex consumer products and packaging at lower costs. This evolution extends to the manufacturing of lighting products, impacting the consumer industry positively.

1.5 GREEN MANUFACTURING IN MATERIALS PROCESSING

Manufacturing entails a conversion process whereby diverse inputs are transformed into products or subcomponents that meet unfulfilled needs. Inputs include raw materials, energy, cutting fluids, and processes, while outputs consist of waste, emissions, and finished items. The industrial sector consumes approximately 37% of global energy production, necessitating the adoption of more efficient approaches and technologies in manufacturing to achieve the vision of a sustainable society (Singh et al., 2018). Sustainable manufacturing seeks to achieve equivalent output with reduced input, thereby lowering overall consumption, environmental impacts, and waste. However, theories and approaches aimed at realizing sustainable manufacturing are often intricate, lacking practical guidelines, and predominantly emphasizing the energy aspect (Narayanan and Das, 2014).

1.5.1 Microwave Processing of Materials

Microwave material processing is becoming more and more popular as a green technology to reduce energy use and boost productivity in traditional industrial processes that reduce carbon dioxide emissions (Kaushal et al., 2019). Microwave heating may shorten the time and lower the temperature required for material processing due to its many benefits over traditional techniques, including its capacity to offer internal heating of substances and quick selective heating (Bhoi et al., 2019). Several methods for processing materials using microwaves have recently been investigated, including metal manufacturing, ceramic sintering, and metal powder. Additionally, it has often been observed that thermal non-equilibrium states during microwave processing can increase chemical reactions and cause fast phase mixing at the iron grain boundary.

Materials processed in a microwave have several benefits for the environment that are consistent with the ideas of green manufacturing. First of all, compared to other methods, microwave heating of materials is particularly energy-efficient since it involves direct energy transfer to the materials. Because of its efficiency, there is a decrease in overall energy usage and carbon footprint. Furthermore, microwave processing frequently calls for lower processing temperatures and shorter processing durations, which results in lower energy consumption and emissions. Additionally, it permits targeted and accurate heating, which can minimize material waste and improve process control, maximizing the use of available resources (Kaushal et al., 2021). Additionally, microwaves might make it easier to recycle and use greener materials by enabling controlled processing of materials that might otherwise be

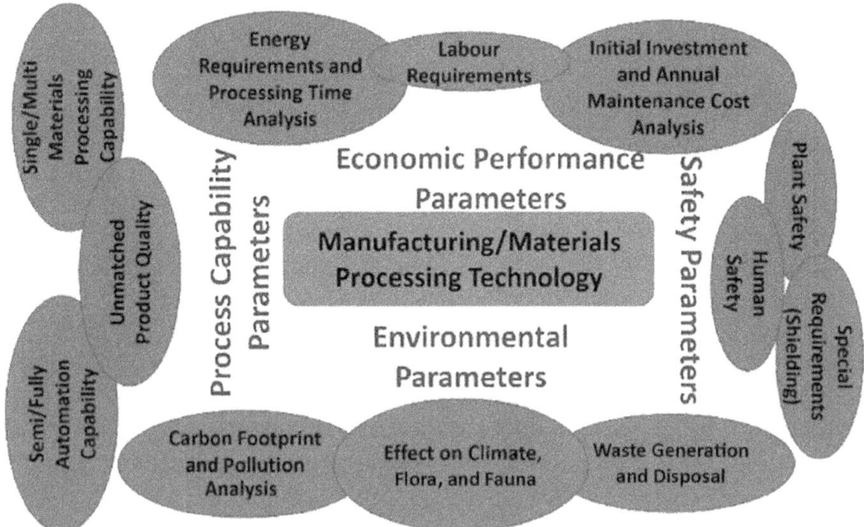

FIGURE 1.4 Some important parameters to study the sustainability of any manufacturing process.

Source: C. Singh et al. (2023).

difficult to manage with traditional methods. When it comes to energy efficiency, lower emissions, improved control, and the possibility of more environmentally friendly material handling, microwave processing is generally considered a greener option.

Numerous aspects of the microwave material processing process need to be compared to traditional systems for it to be viable. Benchmarking performance indicators and then comparing the data to the current system/process is a challenging endeavor. At present, there is no method that can directly deliver the sustainability index for any industrial process in direct contrast to the traditional system. The current study includes a few key factors that may be used to compare the sustainability of processes based on how well they operate and score in the relevant areas. In Figure 1.4, these characteristics are shown.

1.5.2 Friction Stir Welding Process

Sustainable manufacturing tries to minimize or remove dangerous emissions and wastes by saving natural resources via energy-efficient processes with little environmental effect. Welding is a critical industrial technique used to create complicated goods and structures (Nandan et al., 2008). During the welding process, conventional fusion welding (FW) technologies like arc, gas, and laser welding need a lot of heat input (energy), which may lead to the production of different weld flaws including cracking and porosity during the solidification phase. Additionally, traditional FW uses consumables as fillers and emits hazardous gases or radiations, which results in resource waste and negative environmental effects. In summary, we can say that traditional FW is not an appropriate, eco-friendly, or energy-efficient joining procedure (Majeed et al., 2020).

Friction Stir Welding (FSW) is seen to be a promising method for joining materials of comparable or differing thicknesses below the melting point. It does this by overcoming the drawbacks of traditional FW processes including solidification cracking, porosity, hydrogen embitterment, and slag inclusions (Meng et al., 2021). Due to its benefits and uses, FSW has drawn a lot of interest recently from a variety of industries, such as aerospace, automotive, marine ships, locomotives, and so on. The current research is to explore the numerous sustainable features of FSW and how it will provide diverse sectors with a sustainable manufacturing method in the future. Innovative welding methods that are cost-effective, environmentally friendly, sustainable, and process-efficient are becoming more and more in demand.

Because of the aforementioned features, FSW is a recognized green manufacturing method worldwide. The so-called "green technology," or FSW, is a revolutionary solid-state joining method that involves inserting a specially profiled pin into the faying surfaces of a non-consumable spinning tool until the tool shoulder meets the surface of the workpiece. Following a designated dwell period, the welding process is completed by advancing the tool along the joint line at predefined process parameters, including welding speed, rotating speed, tool tilt angle, tool offset, and so on (Tamsir et al., 2012; Drizo and Pegna, 2006). The friction between the tool and the workpiece, the pin and the workpiece, and the plastic deformation of the workpiece material all contribute to the heat needed for joining. A solid-state junction is created behind the tool pin as a consequence of the pin stirring and moving the plastically deformed material from the leading side to the trailing side as the tool advances along the weld line. The forging action of the tool shoulder aids this process. FSW is a cutting-edge welding method with global applications, particularly in the production of lightweight aluminum structures.

Figure 1.5 represents the sustainability approach of the FSW process.

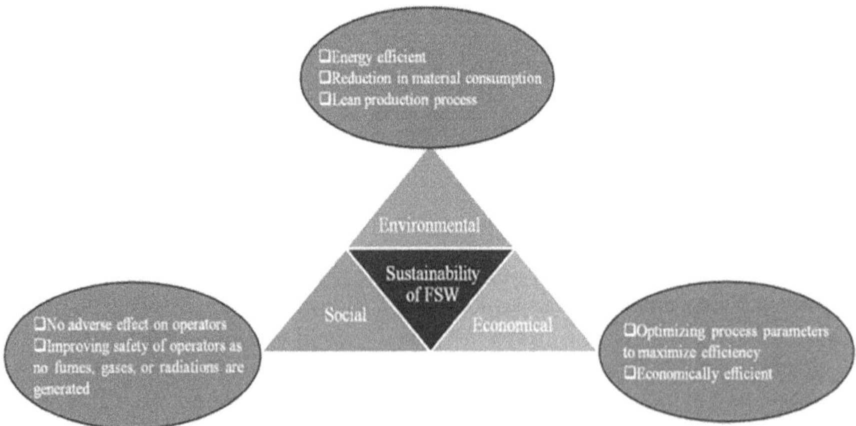

FIGURE 1.5 A diagrammatic representation of the FSW process's sustainability.

Source: Majeed et al. (2021).

1.5.3 Rapid Prototyping Process

The phrase "green manufacturing" refers to a subset of sustainability that takes the environment's influence into account by modifying processes to an acceptable degree. Green manufacturing focuses on the technology and solutions needed to shift from conventional economic practices to sustainable levels of consumption. Through the research and development of technology that eliminates and reduces the use of hazardous chemicals beginning from design, production, and application, green manufacturing should be able to avoid pollution and save energy(Tamsir et al., 2012; Drizo and Pegna, 2006). Reducing energy consumption and material intensiveness throughout the production process is another goal of the green manufacturing method. Manufacturers often focus a lot of their decisions on price, functionality, and quality. Environmental sustainability is also another aspect that has to be taken into account. Eco-efficiency, non-renewable resource concern, and carbon emission reduction are driving forces behind eco-friendly product and process development. One of the technologies that increases the product's eco-efficiency is rapid prototyping (RP).

From a given 3D CAD model, RP often employs additive manufacturing technologies (SLA, SLS, LOM, etc.) to generate a physical model. Because of this, the majority of research in sustainable RP focuses on the tools, methods, and materials utilized in additive manufacturing. A schematic representation of the aspects of the sustainable RP study is shown in Figure 1.6. Figure 1.6 illustrates how the efficiency of 3D CAD modeling systems, data exchange systems, RP equipment, RP processes, process parameter selection systems, and primary materials manufacturing

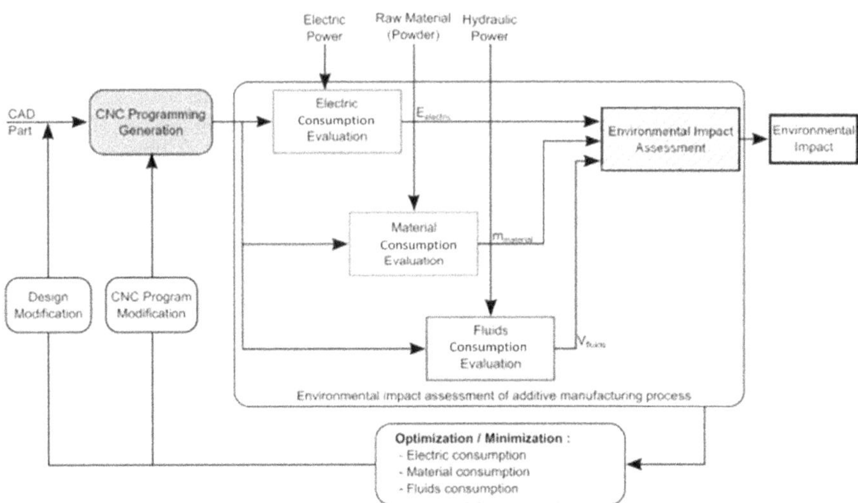

FIGURE 1.6 Global methodology to estimate the environmental impact of CAD part.

Source: Bourhis et al. (2013).

all contribute to the sustainability of RP (Ullah et al., 2013). Numerous writers have looked at the aforementioned aspects and offered their perspectives on how sustainable RP is in terms of greenhouse gas emissions, accuracy, risk, energy, resource, and time usage.

The process of creating tangible items via additive manufacturing is known as RP. Using computer technology, the material is stacked layer by layer to create the specified item from 3D computer models. Rapid software tools and computer numerically controlled machine tools are used in the process. These days, RP is used in the production of a greater variety of products, including high-quality parts in relatively small quantities.

While RP technologies themselves offer these eco-friendly advantages, their overall impact on sustainability depends on various factors, such as the specific materials used, the energy sources powering the machines, and the disposal/recycling methods for unused or waste materials. Nonetheless, these techniques hold promise in aligning with green manufacturing practices as they continue to evolve and incorporate more sustainable approaches.

Certainly! RP, particularly through additive manufacturing methods like 3D printing, embodies several green manufacturing principles such as those discussed in the following subsections.

1.5.3.1 Reduced Material Waste

In conventional manufacturing, the typical process involves subtracting excess material to achieve the desired product shape. Conversely, RP methods, notably 3D printing, follow an additive approach, incrementally layering material to construct the final object. This fundamental difference significantly mitigates material waste as it only utilizes the precise amount needed for the product, eliminating the discard of excess material prevalent in traditional manufacturing (Tamsir et al., 2012). By adding material selectively, layer upon layer, 3D printing minimizes waste generation, ensuring a more efficient use of resources compared to conventional methods, which aligns with sustainable manufacturing practices by reducing material waste and optimizing resource utilization.

1.5.3.2 Energy Efficiency

Additive manufacturing, such as 3D printing, demonstrates superior energy efficiency when contrasted with traditional manufacturing approaches. In contrast to traditional machining or molding techniques, which require significant energy inputs to remove superfluous material from raw materials by cutting or molding, 3D printing functions differently. This novel method uses significantly less energy since it applies material only where it is needed to build the finished product. Through layer-by-layer selective melting, curing, or depositing of material following the digital design, 3D printing reduces energy usage by obviating the needless removal of surplus material. As a result, this focused strategy drastically lowers the total energy needs, establishing additive manufacturing as a more environmentally friendly production option.

1.5.3.3 On-Demand Production

RP, especially with the use of 3D printing, enables localized manufacturing and on-demand production, indicating a move toward more environmentally friendly approaches. This strategy reduces the need for vast transportation networks by enabling the production of commodities closer to their intended destination. RP greatly reduces the carbon footprint usually associated with shipping and transportation in traditional manufacturing models by shortening the time between production and consumption. This decrease in the need for transportation not only reduces emissions but also lessens the need for the surplus inventory that is frequently needed to satisfy the disparate demands across far-off markets. Consequently, RP promotes a greener and more productive production ecosystem by optimizing supply chains, reducing emissions associated with transportation, and reducing the need for large inventories. All of these factors make RP consistent with sustainable principles.

1.5.3.4 Customization and Iteration

RP allows designers and engineers to create and assess prototypes quickly and efficiently. Its primary feature is the ability to digitally iterate designs before beginning physical production. This digital iteration technique drastically eliminates the need to create several physical prototypes. Virtual design refinement reduces physical iterations. As a result, fewer actual prototypes are required, which reduces material consumption and waste throughout the design and testing phases. This streamlined method not only shortens the development cycle but also (Kaushal et al., 2022) contributes to a more resource-efficient process by reducing material usage and waste, which is consistent with sustainable manufacturing standards.

1.5.3.5 Sustainable Materials Usage

The increased emphasis on sustainable materials in 3D printing helps to encourage environmentally responsible production methods. Additive manufacturing has shifted toward using materials that target environmental sustainability. Biodegradable plastics, recyclable materials, and renewable filaments are gaining traction in the additive manufacturing industry. These materials provide a variety of product possibilities while also promoting eco-friendly production processes by lowering reliance on non-renewable resources and limiting the environmental effect of traditional material use. This emphasis on sustainable materials in 3D printing demonstrates a dedication to greener manufacturing techniques, in line with the global push for more ecologically responsible production methods.

1.5.3.6 Small-Batch Production

Additive manufacturing enables cost-effective small-batch production, providing an alternative to large-scale manufacturing runs. This feature reduces the prevalence of overproduction and surplus inventory, which are frequent in traditional manufacturing systems. Additive manufacturing eliminates waste from surplus stock and obsolete inventory by allowing for more precise production that is aligned with demands. This leaner approach not only maximizes resource consumption but also fits with lean manufacturing concepts, stressing efficiency, reduced waste, and the elimination of unneeded inventory, all of which contribute to a more sustainable and resource-conscious manufacturing process.

1.5.3.7 Design Optimization and Lifecycle Assessment

RP accelerates early-stage design optimization, resulting in the development of products with improved resource efficiency, durability, and end-of-life recycling capabilities. This iterative technique allows designers to fine-tune product designs while taking into account environmental impact elements throughout the product's lifecycle. RP produces goods that are more optimized for resource usage, improving longevity while also making recycling or disposal easier after their lifecycle ends. This iterative methodology not only improves the product's overall environmental sustainability but also encourages designers to take a holistic approach, emphasizing careful consideration of ecological factors throughout the product's life cycle.

1.5.3.8 Reduction in Tooling Requirements

Traditional manufacturing relies heavily on specialized molds, tools, and fixtures, which typically require large resources to produce and maintain. In contrast, additive manufacturing significantly reduces the need for specific tools by creating components layer by layer from digital designs. This reduction in tooling requirements not only saves materials but also reduces waste generated during tool production and maintenance. The capacity of additive manufacturing to generate detailed and personalized parts without the need for elaborate tooling not only streamlines the production process but also significantly contributes to resource conservation and waste reduction, which is consistent with sustainable manufacturing principles.

RP fits with green manufacturing by minimizing waste, energy consumption, and transportation needs while also allowing for sustainable material utilization and localized production, all of which contribute to a more environmentally friendly manufacturing process.

1.5.4 Sustainable Grinding Processes

The primary goal of this part is to introduce sustainable manufacturing methods in machining that aim to reduce the negative environmental impacts of manufacturing processes.

Grinding is a regularly used manufacturing procedure for producing final components with precise shape, size, and accuracy. It employs multiple point-cutting tools to eliminate unwanted materials from the stock in the form of chips. Previously, scholars have categorized grinding into two main types: Form-Finish Grinding (FFG) and Stock Removal Grinding (SRG). FFG encompasses surface grinding, internal grinding, and cylindrical grinding operations, while SRG involves snagging and cut-off procedures. The grinding process holds significant importance in the manufacturing industries, contributing substantially to the production or fabrication of various products. Studies indicate that among manufacturing sectors, grinding holds the primary position with a 25% share, followed by turning, milling, and drilling.

1.5.4.1 Ultrasonic-Assisted Grinding

Grinding stands as the predominant technique for machining materials such as ceramics, composites, alloys, and super alloys distinguished by their high strength-to-weight ratios. Nonetheless, the conventional grinding method encounters numerous challenges attributable to the exacting mechanical properties of these advanced materials. These challenges encompass environmental concerns, worker health risks, preservation of surface integrity, material stability, and heightened energy demands throughout the machining operation (Singh et al., 2020).

1.5.4.2 Cryogenic Grinding

Cryogenic, derived from the Greek word "cryo," meaning cold, pertains to the study of low temperatures and the behavior of materials under such conditions. In this process, grinding is conducted using liquid nitrogen as a cryogen for the work materials. Cryogenic machining aims to lower the temperature of the cutting zone relative to the surroundings by employing an inert gas. This mechanism facilitates efficient heat transfer from the machining zone, thereby reducing the formation of thermal stresses in the work material, resulting in enhanced surface integrity. Research into cryogenic grinding has demonstrated a notable decrease in grinding forces and specific energy consumption when compared to conventional flood grinding and dry grinding methods. Additionally, the effectiveness of cryogenic cooling was observed to increase with the ductility of the work material (Paul et al., 1993).

The utilization of liquid nitrogen as a cryogen has demonstrated reductions in grinding force by up to 24% and 12% compared to dry and wet grinding environments, respectively. Similarly, a decrease in surface roughness ranging from 38% to 15% has been observed in cryogenic environments when compared to dry and wet environments, respectively, when using Sol-gel grinding wheels. The cryogenic environment has been observed to induce residual tensile stresses aligned with the direction of grinding. In contrast, residual compressive stresses perpendicular to the grinding direction have increased by 5%. When using cryogenic lubrication settings, the increase in residual compressive stresses contributes to increased surface integrity, wear resistance, and reduced tensile residual stress.

1.6 CONCLUSION

Green manufacturing prioritizes improving production practices over regulating technology to conserve natural resources for future generations and recycle materials. According to several researchers, it is good to use it for economic advancement while concurrently reducing resource depletion, waste output, and all forms of pollution. The study encourages the use of green technology over traditional techniques because of its environmentally friendly design, less waste, and better customer satisfaction. Using traditional ways wastes natural resources and pollutes our environment. Sustainable manufacturing aims at the reduction or elimination of harmful gases, and wastes, and optimum utilization of natural resources with energy-efficient processing involving minimum impact on the environment. Manufacturers may reduce their environmental impact and remain competitive in a fast-changing global market through the use of sustainable design concepts, lean manufacturing processes, energy efficiency measures, waste reduction, and cutting-edge technology. Implementing these strategies is a responsible choice and a strategic advantage in the 21st century.

REFERENCES

Bhoi, N. K., Singh, H., Pratap, S., & Jain, P. K. (2019). Microwave material processing: a clean, green, and sustainable approach. In *Sustainable Engineering Products and Manufacturing Technologies* (pp. 3–23). Elsevier. https://doi.org/10.1016/B978-0-12-816564-5.00001-3

Bourhis, F. Le, Kerbrat, O., Hascoet, J.-Y., & Mognol, P. (2013). Sustainable manufacturing: evaluation and modeling of environmental impacts in additive manufacturing. *The International Journal of Advanced Manufacturing Technology*, *69*(9), 1927–1939. https://doi.org/10.1007/s00170-013-5151-2

Despeisse, M., Chari, A., González Chávez, C. A., Monteiro, H., Machado, C. G., & Johansson, B. (2022). A systematic review of empirical studies on green manufacturing: eight propositions and a research framework for digitalized sustainable manufacturing. *Production and Manufacturing Research*, *10*(1), 727–759. https://doi.org/10.1080/21693277.2022.2127428

Dilip Maruthi, G., & Rashmi, R. (2015). Green manufacturing: it's tools and techniques that can be implemented in manufacturing sectors. *Materials Today: Proceedings*, *2*(4–5), 3350–3355. https://doi.org/10.1016/j.matpr.2015.07.308

Drizo, A., & Pegna, J. (2006). Environmental impacts of rapid prototyping: an overview of research to date. *Rapid Prototyping Journal*, *12*(2), 64–71. https://doi.org/10.1108/13552540610652393

Ermolaeva, N. S., Castro, M. B. G., & Kandachar, P. V. (2004). Materials selection for an automotive structure by integrating structural optimization with environmental impact assessment. *Materials and Design*, *25*(8), 689–698. https://doi.org/10.1016/j.matdes.2004.02.021

Haleem, A., Javaid, M., Singh, R. P., Suman, R., & Qadri, M. A. (2023). A pervasive study on Green Manufacturing towards attaining sustainability. *Green Technologies and Sustainability*, *1*(2), 100018. https://doi.org/10.1016/j.grets.2023.100018

Holloway, L. (1998). Materials selection for optimal environmental impact in mechanical design. *Materials and Design*, *19*, 133–143.

Kaushal, S., Gupta, D., & Bhowmick, H. (2019). On processing and flexural behaviour of functionally graded clads developed through microwave irradiation. *Materials Research Express*, *6*. https://doi.org/10.1088/2053-1591/ab11ea(Not accessible as of [2024/06/14])

Kaushal, S., Gupta, D., & Bhowmick, H. (2021). Wear behavior of microwave-processed Ni-WC8Co-based functionally graded materials. *Proceedings of the Institution of Mechanical Engineers, Part L: Journal of Materials: Design and Applications*, *235*, 146442072098811. https://doi.org/10.1177/1464420720988119

Kaushal, S., Singh, I., Singh, S., & Gupta, A. (2022). *Sustainable Advanced Manufacturing and Materials Processing*. CRC Press. https://doi.org/10.1201/9781003269298

Leong, W. D., Lam, H. L., Ng, W. P. Q., Lim, C. H., Tan, C. P., & Ponnambalam, S. G. (2019). Lean and green manufacturing—a review on its applications and impacts. In *Process Integration and Optimization for Sustainability* (Vol. 3, Issue 1, pp. 5–23). Springer. https://doi.org/10.1007/s41660-019-00082-x

Lin, G., & Hao, B. (2020). Research on green manufacturing technology. *Journal of Physics: Conference Series*, *1601*(4). https://doi.org/10.1088/1742-6596/1601/4/042046

Ljungberg, L. Y. (2007). Materials selection and design for development of sustainable products. *Materials & Design*, *28*, 466–479. https://doi.org/10.1016/j.matdes.2005.09.006

Majeed, T., Wahid, M. A., Alam, M. N., Mehta, Y., & Siddiquee, A. N. (2020). Friction stir welding: a sustainable manufacturing process. *Materials Today: Proceedings*, *46*, 6558–6563. https://doi.org/10.1016/j.matpr.2021.04.025

Majeed, T., Wahid, M., Alam, N., Mehta, Y., & Siddiquee, A. N. (2021). Friction stir welding: a sustainable manufacturing process. *Materials Today: Proceedings*. https://doi.org/10.1016/J.MATPR.2021.04.025

Meng, X., Huang, Y., Cao, J., Shen, J., & Santos, J. F. dos. (2021). Recent progress on control strategies for inherent issues in friction stir welding. *Progress in Materials Science*. https://doi.org/10.1016/J.PMATSCI.2020.100706

Miranda, I. T. P., Moletta, J., Pedroso, B., Pilatti, L. A., & Picinin, C. T. (2021). A review on green technology practices at BRICS countries: Brazil, Russia, India, China, and South Africa. *SAGE Open*, *11*(2). https://doi.org/10.1177/21582440211013780

Nandan, R., DebRoy, T., & Bhadeshia, H. K. D. H. (2008). Recent advances in friction-stir welding: process, weldment structure and properties. *Progress in Materials Science*. https://doi.org/10.1016/J.PMATSCI.2008.05.001

Narayanan, R. G., & Das, S. (2014). Sustainable and green manufacturing and materials design through computations. *Proceedings of the Institution of Mechanical Engineers, Part C: Journal of Mechanical Engineering Science*, *228*(9), 1581–1605. https://doi.org/10.1177/0954406213508754

Paul, I. D., Bhole, G. P., & Chaudhari, J. R. (2014). A review on green manufacturing: it's important, methodology and its application. *Procedia Materials Science*, *6*, 1644–1649. https://doi.org/10.1016/j.mspro.2014.07.149

Paul, S., Bandyopadhyay, P. P., & Chattopadhyay, A. B. (1993). Effects of cryo-cooling in grinding steels. *Journal of Materials Processing Technology*, *37*(1), 791–800. https://doi.org/https://doi.org/10.1016/0924-0136(93)90137-U

Singh, A., Philip, D., Ramkumar, J., & Das, M. (2018). A simulation based approach to realize green factory from unit green manufacturing processes. *Journal of Cleaner Production*, *182*, 67–81. https://doi.org/10.1016/j.jclepro.2018.02.025

Singh, A. K., Kumar, A., Sharma, V., & Kala, P. (2020). Sustainable techniques in grinding: state of the art review. *Journal of Cleaner Production*, 269. https://doi.org/10.1016/j.jclepro.2020.121876

Singh, C., Khanna, V., & Singh, S. (2023). Sustainability of microwave heating in materials processing technologies. *Materials Today: Proceedings*, *73*, 241–248. https://doi.org/https://doi.org/10.1016/j.matpr.2022.07.216

Stoffels, P., Kaspar, J., Baehre, D., & Vielhaber, M. (2017). Holistic material selection approach for more sustainable products. *Procedia Manufacturing*, *8*, 401–408. https://doi.org/10.1016/j.promfg.2017.02.051

Tamsir, A., Irawan, R., & Muhida, R. (2012). Rapid prototyping and evaluation for green manufacturing. *International Journal of Engineering Science and Technology Development*, *1*(1), ISSN 2337-3180(Online) ©Universitas Bandar Lampung 2013.

Tao, P., & Zhao, G. (2016). Research on the green manufacturing system and its structure. *Proceedings of the 6th International Conference on Electronic, Mechanical, Information and Management Society*. https://doi.org/10.2991/emim-16.2016.58

Ullah, A. M. M. S., Hashimoto, H., Kubo, A., & Tamaki, J. (2013). Sustainability analysis of rapid prototyping: material/resource and process perspectives. *International Journal of Sustainable Manufacturing*, *3*(1), 20–36. https://doi.org/10.1504/IJSM.2013.058640

Zarandi, M. H. F., Mansour, S., Hosseinijou, S. A., & Avazbeigi, M. (2011). A material selection methodology and expert system for sustainable product design. *International Journal of Advanced Manufacturing Technology*, *57*(9–12), 885–903. https://doi.org/10.1007/s00170-011-3362-y

2 Microwave Processing of Materials

A Sustainable Manufacturing Approach

Yudhveer Kumar Verma, Kanwarjeet Singh, Sandeep Bansal, and Sarbjeet Kaushal

2.1 INTRODUCTION

For over five decades, a diverse group of scientists, including material scientists, physicists, chemists, and engineers, have been exploring the potential of harnessing microwave (MW) energy's interaction with materials (Varghese et al. 2014). Microwaves as a heating source have emerged as a recent trend in manufacturing technology. While numerous heating sources have been used in manufacturing, microwave heating represents a newly developed method in this field. Microwaves possess specific electric and magnetic properties that operate within a certain frequency range over time (Decareau et al. 1986). The significance of microwave processing technology lies in its advantages, particularly its ability to provide uniform heating at the molecular level. This aspect has garnered tremendous importance in the modern manufacturing world. Efficiency and cost-effectiveness play a crucial role in driving the adoption of microwave technology (James et al. 2005). The use of dielectric heating in various industries has added to its economic and efficiency benefits. One of the key advantages of microwave processing is its capability to achieve uniform and homogeneous heating, leading to reduced power consumption and enhanced efficiency. Continuous advancements in microwave technology further reinforce its position for future decades, making products more reliable and energy-efficient compared to other methods (Schiffmann et al. 2001). In the realm of manufacturing technology, various techniques have been employed for different applications, such as drying, firing, curing, vulcanization, and thermal food processing. While ohmic heating is utilized in thermal food processing, microwave heating presents its advantages for several applications including baking, cooking, thawing, tempering, drying, pasteurization, sterilization, blanching, and waste treatment (Rosenberg et al. 1987; Anantheswaran et al. 2001; Oda et al. 1992; Sahota et al. 2021). The electrical furnace has also been a prominent tool in manufacturing technology. However, microwave heating stands out due to its unique performance characteristics, including penetration radiation,

DOI: 10.1201/9781003449225-2

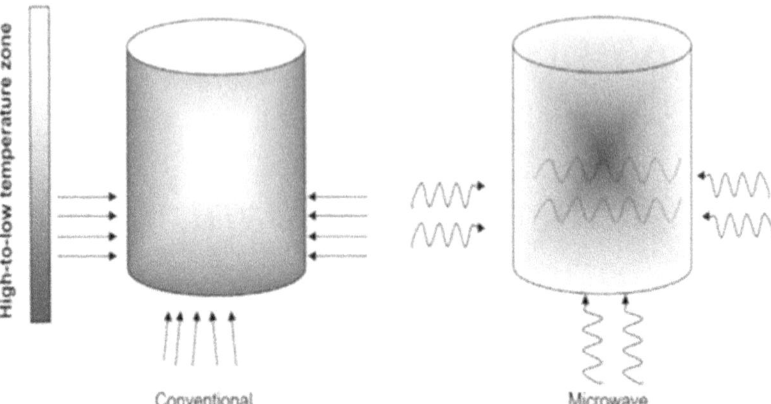

FIGURE 2.1 Schematic diagram of conventional and microwave heating pattern.

controllable electric field distribution, rapid and selective heating, and the ability to self-limit reactions. These features set microwave processing apart from conventional heating sources, making it a compelling choice for various applications (Srinath et al. 2011; Pal et al. 2020). The microwave has many modern applications. It has found use in the process of extracting plant parts for medicinal purposes, known as natural metabolites (Singh et al. 2015). Microwave processing is used in the food business as a heating source in numerous applications (Tamang et al. 2021). The use of microwave technology has expanded to include the chemical industry and other manufacturing sectors. Microwave-assisted extraction is used in some applications to produce the products needed for material processing. Biofuels, cellulose, oil, resins, and pharmaceuticals are some examples. It can be used to dry materials including powder, foam, wood, and cloth (Baghel et al. 2023). The use of microwaves in radar, antennas, radio, and especially 3D printing and additive manufacturing (Kumar et al. 2015). Figure 2.1 shows a schematic diagram of conventional and microwave heating patterns.

2.2 USE OF MICROWAVE AS HEATING SOURCE

In the industrial world, various electromagnetic radiations, including gamma rays, X-rays, ultraviolet rays, infrared rays, and radio waves, find applications in different technologies. Among these radiations, microwaves stand out due to their versatile use in both traditional and modern technologies. Microwaves have wavelengths ranging from 1 mm to 1 m and are considered a type of radio wave, falling within the electromagnetic spectrum. The electromagnetic spectrum comprises a wide range of frequencies, and these radiations are classified based on their frequency and wavelength. This classification is vital as it determines their specific applications, not just in manufacturing but also in various information technology fields. Regardless of their frequency and wavelength, all these waves are collectively known as electromagnetic (EM) waves, and they all travel at the speed of light. The use of microwaves is particularly significant in industrial applications due to their unique properties and benefits, including controlled and selective heating, efficient communication, and non-destructive testing. Their versatility extends beyond traditional applications to emerging technologies, making them an indispensable part of the modern industrial landscape.

2.3 JOINING

2.3.1 METAL JOINING

Indeed, fusion technology has long been the conventional method for joining metals, but recently, microwave technology has emerged as an innovative approach in the field. Researchers have conducted numerous studies exploring the potential of microwave joining for various metals. During the microwave joining process, intermetallic compounds like carbide and cementite are formed at the joint interfaces. One of the significant advantages of microwave technology is the uniform heating it provides, resulting in a slower cooling rate. This controlled cooling rate allows gases to escape effectively, preventing the formation of porosity and other defects in the joint. Moreover, the uniform and homogeneous heating associated with microwave hybrid heating leads to the formation of a cellular and uniform structure in the joined metals. This outcome showcases the effectiveness of microwave technology in achieving superior joint properties compared to traditional fusion methods. Overall, microwave technology has proven to be a promising alternative in the metal joining process, offering distinct advantages in terms of defect reduction and improved joint structure. As further research and advancements are made, microwave joining is expected to gain even more prominence in industrial applications (Dutta et al. 2017; Pal et al. 2018; Bagha et al. 2016, 2019; Soni et al. 2018; Singh et al. 2019). Researchers have already tried connecting ferrous metals. The joining was accomplished with success. Recent experiments have brought newer material into the mix. Non-ferrous metals must be successfully joined in order to be used in the more modern industrial environment. Figure 2.2 shows a schematic diagram of the microwave joining process.

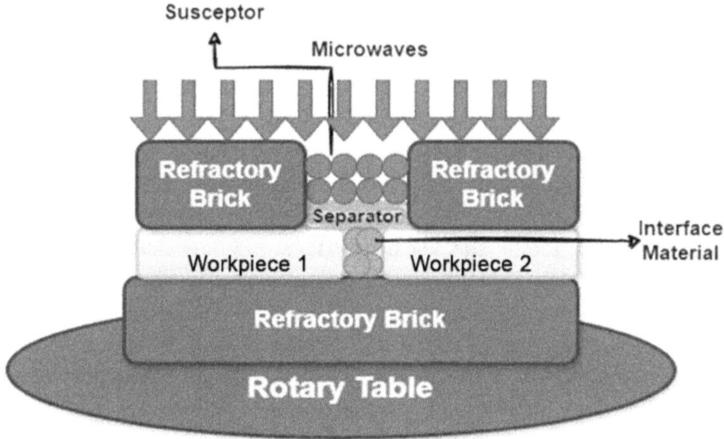

FIGURE 2.2 Figure shows a schematic diagram of microwave joining process.

2.3.2 Ceramic Joining

The increased application of Metal Matrix Composites (MMC) and ceramics in aerospace and industrial settings necessitates suitable joining technologies to meet the current demands. It is crucial that these joining methods preserve the microstructure and potentially enhance material properties. Microwave processing technology emerges as a promising solution in this context. The unique advantage of microwave processing lies in its ability to achieve selective heating at the molecular level. This results in a more uniform and homogeneous microstructure, without any heat-affected zone. Nonmetals have been playing a significant role in the manufacturing industry for quite some time, with ceramics and Fiber-Reinforced Polymer Composites (FRSP) and MMC being widely used. However, joining ceramics and MMC presents a considerable challenge, as traditional welding processes prove inefficient for these materials. The rapid growth of microwave technology has extended its application beyond metals and found extensive use in joining nonmetals like ceramics and MMC. This is primarily due to its capability to selectively and volumetrically heat the materials at the molecular level, leading to a more homogeneous and refined grain structure. The improved homogeneity and refinement in the grain structure not only enhance mechanical properties but also contribute to improved chemical and physical properties. As researchers delve into the realm of nonmetals, especially ceramics, they explore various parameters to optimize and enhance material properties. By leveraging the benefits of microwave technology, researchers aim to achieve superior material joining and further advance the application of MMC and ceramics in aerospace and industrial sectors. With continuous research and development, microwave processing is expected to revolutionize the field of nonmetal joining, opening up new possibilities for innovative manufacturing applications (Gupta et al. 2015; Bansal et al. 2012; Lingappa et al. 2017; Singh et al. 2016, 2018; Srinath et al. 2012; Badiger et al. 2015; Silberglitt et al. 1993; Shukla et al. 2016; Rosa et al. 2013; Siores et al. 1995; Kaushal et al. 2019, 2022a, 2023; and Varadan et al. 1991).

2.3.3 Thermoplastics Joining

Thermoplastics have become an integral part of modern life, finding extensive use in various applications, ranging from sports equipment and toys to CDs, DVDs, food containers, and more. Despite their widespread use, joining thermoplastics poses a significant challenge. While metals can be relatively easily joined through fusion welding, thermoplastics require more complex processes. Researchers worldwide have been exploring fusion welding techniques for thermoplastics, where the joint interface is melted, reducing the viscosity of the polymer and allowing for intermolecular diffusion. This diffusion enables the polymer chains to intertwine across the joint interface, providing strength to the joints. However, this fusion bonding may not always be sufficient, especially if the polymer chains in the joint interface do not have enough fusion. This can result in the polymer remaining intact at that position, leading to the induction of residual stresses when heat is generated during welding. Various factors such as heat input, cooling rates, and the volume fraction of fibers and matrix material can influence the residual stress. Another challenge lies in the orientation of polymer fibers. During the molten state of welding, the fibers can reorient themselves in a way that reduces the strength of the weld. To overcome these challenges, researchers have explored various processes, and one of the most promising methods is microwave welding.

Microwave welding utilizes microwaves as a heating source, offering homogenized and uniform volumetric heating through hybrid heating. This technology melts and welds materials at the molecular level. The advantages of microwave welding include higher heating rates, noncontact heating, high efficiency in generation and conduction, very low power consumption, and ease of heating control. Initially used primarily to heat food in microwave ovens, further innovation led to its application in welding both metals and nonmetals. For nonmetals, such as thermoplastics, microwave welding has been employed to achieve successful joins. Although microwave welding of thermoplastics has been in development for some time, significant advancements occurred in 1993 when TWI (The Welding Institute) established facilities to explore the feasibility and optimize the process for industrial applications. As technology continues to progress, microwave welding is expected to play an increasingly crucial role in joining thermoplastics and other materials, offering efficient and effective solutions for diverse industrial applications (Sosa et al. 2016; Wise et al. 2001; Foong et al. 2021; Wu et al. 1997; Ku et al. 1999; Yarlagadda et al. 1998; Potente et al. 2003; Barasinski et al. 2018; Savu et al. 2016; Pan et al. 2013; Sun et al. 2018, 2019; Singh et al. 2015; Samyal et al. 2020a; Mishra et al. 2020; Mishra et al. 2019; Perumalla et al. 2020; Hong et al. 2020; Rosa 2013; Kazemi et al. 2019).

2.4 MICROWAVE CLADDING

Metal cladding or coating is a type of protective layer where a material, often metal powder, is bonded to a substrate using heat or pressure. The primary purpose of metal cladding is to provide protection against corrosion and reduce material wear. Additionally, cladding helps in reducing surface distortion, roughness, and waviness by compensating for nonuniformities on the metal surface, thus enhancing its overall quality. The crucial aspect of cladding is the bonding of the metal powder with the substrate surface. Various methods have been developed for this purpose, but what is particularly important is achieving selective and volumetric heating to ensure proper bonding. This is where microwave technology proves to be highly effective. Microwave technology enables selective and uniform heating of the metal powder, allowing it to stick to the substrate at the molecular level. The microwave radiations serve as a heat source, facilitating the bonding process between the metal powder and the substrate.

By utilizing microwave technology in the cladding process, manufacturers can achieve a more efficient and precise bonding of the protective layer, leading to enhanced material durability and improved performance. As technology continues

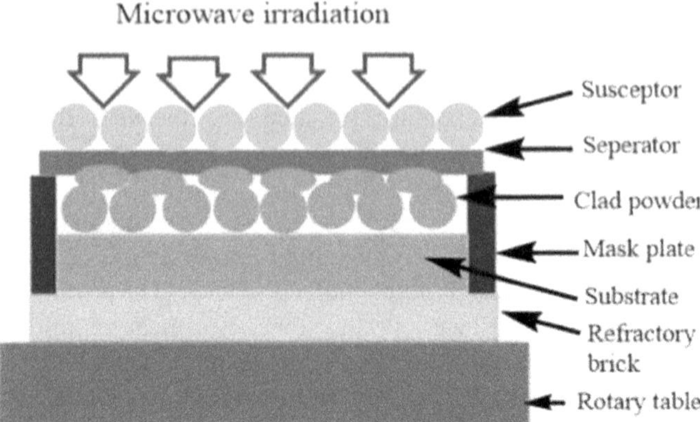

FIGURE 2.3 Schematic diagram of microwave material heating.

to advance, microwave-based metal cladding is expected to gain even more prominence as a reliable and cost-effective protective coating method in various industries (Kumar et al. 2018, Bansal et al. 2022b, 2023, Kaushal et al. 2021, Saloni et al. 2022, Kaushal et al. 2022). Figure 2.3 shows a schematic diagram of microwave material heating in microwave cladding process.

2.5 CASTING

The process of heating and melting bulk metals using microwave radiation requires specialized equipment. The setup typically includes a ceramic crucible that is microwave-absorbent, an insulated thermal casket that is transparent to microwaves, and a multimode microwave cavity. The metal charge is placed inside the open ceramic crucible, which is then covered with the insulated casket. The entire crucible assembly is then positioned within the high-power multimode microwave cavity. The microwave cavity is designed to evenly and homogeneously heat the crucible to the required temperature (Aliakbari et al. 2018). In this setup, the ceramic crucible plays a crucial role in absorbing the microwave energy directed toward the cavity. As a result of three different heat transfer phenomena—radiation, conduction, and convection—the walls of the crucible are heated, which in turn heats the metal charge in close proximity to the walls (Chai et al. 2016). The thermally insulating casket contributes to the increased efficiency of heat generation by trapping the heat inside the crucible, preventing its loss to the surroundings. This heating method enables the effective melting of metal objects that cannot be directly heated by microwave energy. By using microwave radiations in this controlled and targeted manner, metal melting can be achieved with improved efficiency and accuracy. This technology is valuable in various industries where controlled and precise melting of bulk metals is required for manufacturing and processing purposes (Leo et al. 2016; Samyal et al. 2020b; Bansal et al. 2014).

2.6 HEAT TREATMENT

Microwave technology finds valuable application in the heat-treatment process, enhancing the properties of materials. The heat treatment is accomplished using a microwave furnace, and three different techniques are commonly utilized (Pal et al. 2018; Chandrasekaran et al. 2018; Singh et al. 2015; Kumar et al. 2020).

1. Molten salt bath processing: In this method, the substrate is immersed in various salt liquids, such as nitrate, carbonate, chloride, and caustic solutions. The heat treatment is then carried out with precise control over concentration, time, and temperature. The molten salt bath ensures uniform heating and controlled diffusion of elements within the material, resulting in desired material properties.
2. Granular suscepting media: This method involves placing the substrate within a granular material that efficiently absorbs microwaves. The granular material also aids in neutralizing any nonuniformity in the electric and magnetic fields, ensuring homogenized heating of the part. The granular suscepting media facilitates quick and efficient heating of the part, leading to improved heat-treatment results.
3. Fluidized-bed processing: In this technique, particles are suspended in a gaseous stream, behaving like a fluid. The fluidization process is achieved by floating gases like LPG, argon, natural gas, ammonia, and nitrogen into the atmosphere. These gases interact with the heated surface of the material and are absorbed by it, resulting in the desired mechanical and metallurgical properties in the metal parts.

Each of these heat-treatment techniques using microwave technology offers advantages in terms of uniformity, efficiency, and controlled heating. Microwave heat treatment provides opportunities for precise and tailored processing, leading to enhanced material properties and improved performance in various industrial applications. The microwave technology not only enhances the grain's structure but also improves the dexterity of alloys as uniform volumetric heating takes place which refines the microstructure and improves the mechanical properties (Kumar et al. 2020 and Bansal et al. 2023).

2.7 MICROWAVE SINTERING

Sintering is a crucial process that significantly impacts the microstructure and mechanical properties of materials. During sintering, loose or powdered materials are compressed and heated at high temperatures and pressure to create a compact and solid piece. Microwave sintering is a modern method that utilizes a furnace with a controlled atmosphere at 2.45 GHz and multimode equipment. The sintering process occurs within a temperature range of 110°C to 1300°C, with soaking times ranging from 5 to 60 minutes (Kumar et al. 2021). Compared to conventional sintering methods, microwave sintering offers distinct advantages. The selective and homogeneous heating of materials during microwave sintering leads to improved mechanical properties in the resulting samples. Essential mechanical properties, such as hardness, yield strength, and modulus of rupture, were found to be higher in materials processed using microwave sintering. This can be attributed to the molecular-level heating of the powder, which causes the particles to fuse together, forming a cohesive substance. Subsequently, applying pressure during the compaction process binds the powder together to create a dense and well-formed tablet or material. Numerous researchers have extensively studied the microwave sintering process, and their findings demonstrate that it not only enhances mechanical and microstructural properties but also improves chemical and physical properties. The controlled and efficient heating achieved through microwave sintering contributes to the development of materials with superior performance characteristics, making it a promising technique for various industrial applications. As research in this field continues, microwave sintering is expected to play an increasingly significant role in advancing materials engineering and manufacturing processes.

2.8 MACHINING

Microwave machining is a relatively newer manufacturing technique developed for cutting materials into desired shapes and sizes. The process involves using microwaves for cutting, which minimizes damage caused by intensive heating and discharge operations. One of the areas where microwave machining has been studied extensively is in the machining of dielectrics. In dielectrics machining with microwave technology, a thermal pulse is generated using a laser, and it is focused on a small area where localized heating occurs. This focal point serves as the site of microwave absorption. Researchers have conducted various experiments in this field, and the results have shown that with microwave machining, the volume of material removed can be improved by at least eight times per minute compared to conventional machining methods. The advantages of microwave machining lie in its ability to provide precise and localized heating, which leads to enhanced material removal rates and reduced damage to the workpiece. Additionally, microwave machining has the potential to improve process efficiency and reduce machining times, making it an attractive option for various manufacturing applications, especially in the machining of dielectrics and other materials that are sensitive to excessive heat or mechanical stress. As further research and development take place, microwave machining is expected to gain more prominence as a viable and efficient manufacturing method (Babu et al. 2022 and Kaushal et al. 2018).

2.9 DRILLING

Microwave-assisted drilling is a technique that enhances the drilling process by achieving more robust and cleaner results, eliminating chips and burrs commonly associated with conventional drilling methods. This method involves the use of a microwave concentrator, which converts microwave energy into a near-field around the concentrator. The drilling process begins with placing the concentrator at rest on the surface of the workpiece using a fixture. An applied electromagnetic field induces an electric field, which is neutralized by the free electrons present in the concentrator. This arrangement allows the surface charge density to increase at the end of the concentrator, resulting in the drilling of a hole. As the electric field rises above the ionization potential of the dielectrics surrounding the concentrator, ionization of the adjacent area occurs, leading to the generation of plasma. The plasma takes on a spherical form and interacts with the material around the concentrator, which has been precisely focused on the area to be drilled. The drilling hole is obtained through the process of ablation and evaporation, where the material is removed by the action of the generated plasma (Chandrasekaran et al. 2011). The advantage of microwave-assisted drilling lies in its ability to produce cleaner and more precise holes without the typical chips and burrs seen in conventional drilling. The controlled application of microwave energy and plasma generation ensures a cleaner drilling process with reduced material damage. This technique is particularly useful in applications where precision and cleanliness are essential, such as in the aerospace and electronics industries. As microwave-assisted drilling technology advances, it is expected to find broader applications and further improve drilling efficiency and accuracy.

Microwave radiations are used in the process, which results in heating, melting, and ultimately ablation and material evaporation. By selectively heating at the molecular level a concentrated region created by microwaves, the specimen is heated and melted in such a way that the areas around are not impacted and exact control of heat is occurring.

2.10 CONCLUSION

- Microwave processing is well-liked because it has advantages such as energy merit, shortened cycle time, cost and power savings, and enhanced physical, chemical, and mechanical properties of materials.
- The concentration of electric and magnetic fields on the material causes a rise in surface charge density, which ionizes the space between them and heats the material. This kind is applied to a variety of substrates, such as metals, nonferrous metals and alloys, and nonmetals, such as composites and ceramics.
- A novel approach to material joining has been created, and future observation shows promising results.
- A novel method for improving surface morphology by microwave coating or cladding has been presented.
- Common materials' heating, melting, evaporation, and ablation were taken into account, and several recent industrial uses of microwaves were reported in the review.
- The new cutting technology that might be developed, such as plasma arc, laser, and microwave cutting, is the subject of further research.

REFERENCES

Aliakbari, S., Ketabchi, M., & Mirsalehi, S. E. (2018). Through-thickness friction stir processing; a low-cost technique for fusion welds repair and modification in AA6061 alloy. *Journal of Manufacturing Processes*, 35, 226–232.

Anantheswaran, R. C., & Ramaswamy, H. S. (2001). Bacterial destruction and enzyme inactivation during microwave heating. In *Handbook of Microwave Technology for Food Application* (pp. 191–214). CRC Press.

Babu, A., Arora, H. S., Singh, R. P., & Grewal, H. S. (2022). Slurry erosion resistance of microwave derived Ni-SiC composite claddings. *Silicon*, 14, 1–13.

Badiger, R. I., Narendranath, S., & Srinath, M. S. (2015). Joining of Inconel-625 alloy through microwave hybrid heating and its characterization. *Journal of Manufacturing Processes*, 18, 117–123.

Bagha, L., Sehgal, S., & Thakur, A. (2016). Comparative analysis of microwave based joining/welding of SS304-SS304 using different interfacing materials. In MATEC Web of Conferences (Vol. 57, p. 03001). EDP Sciences.

Bagha, L., Sehgal, S., Thakur, A., Kumar, H., & Goyal, D. (2019). Low cost joining of SS304-SS304 through microwave hybrid heating without filler-powder. *Engineering Research Express*, 1(2), 025035.

Baghel, P. K. (2023). Application of microwave in manufacturing technology: A review. *Materials Today: Proceedings*. https://doi.org/10.1016/j.matpr.2023.02.008

Bansal, S., Kaushal, S., Gupta, D., & Jain, V. (2022a). On microstructure and cavitation erosion behavior of microwave-synthesized Ni-Al_2O_3-based composite claddings, *Surface Review and Letters*, 2240006.

Bansal, S., Kaushal, S., Mago, J., Gupta, D., Jain, V., Babbar, A., & Sharma, D. (2022b). Effect of variation of WC reinforcement on metallurgical and cavitation erosion behavior of microwave processed NiCrSiC-WC composites clads. *Proceedings of the Institution of Mechanical Engineers, Part C: Journal of Mechanical Engineering Science*, 237(22), 5460–5475.

Bansal, S., Kaushal, S., Gupta, D., & Jain, V. (2023). Effect of variation of ceramic reinforcement content on metallurgical and cavitation erosion behaviour of microwave processed composite clads, *International Journal of Surface Science and Engineering*, 17(2), 1–14.

Bansal, A., Sharma, A. K., Kumar, P., & Das, S. (2012). Joining of mild steel plates using microwave energy. *Advanced Materials Research*, 585, 465–469.

Bansal, A., Sharma, A. K., Kumar, P., & Das, S. (2014). Characterization of bulk stainless steel joints developed through microwave hybrid heating. *Materials Characterization*, 91, 34–41.

Barasinski, A., Tertrais, H., Bechtel, S., & Chinesta, F. (2018, May). Microwave heating for thermoplastic composites-Could the technology be used for welding applications? In AIP Conference Proceedings (Vol. 1960, No. 1). AIP Publishing.

Chai, X., Yuan, T., & Kou, S. (2016). Liquation and liquation cracking in partially melted zones of magnesium welds. *AWS Welding Journal*, 95(2), 57s–67s.

Chandrasekaran, N. I., Muthukumar, H., Sekar, A. D., Pugazhendhi, A., & Manickam, M. (2018). High-performance asymmetric supercapacitor from nanostructured tin nickel sulfide (SnNi2S4) synthesized via microwave-assisted technique. *Journal of Molecular Liquids*, 266, 649–657.

Chandrasekaran, S., Basak, T., & Ramanathan, S. (2011). Experimental and theoretical investigation on microwave melting of metals. *Journal of Materials Processing Technology*, 211(3), 482–487.

Decareau, R. V. (1986). Microwave food processing equipment throughout the world. *Food Technology* (USA), 40(6).

Dutta, T., Sanwaria, S. J., Vanesh, S., & Dhinakaran, S. P. (2017). Analysis of microwave welding of stainless steel. *International Journal of Scientific & Engineering Research (IJSER)*, 5(4), 92–95.

Foong, P. Y., Voon, C. H., Lim, B. Y., Arshad, M. K. M., Gopinath, S. C., Foo, K. L., Rahim, R. A., & Hashim, U. (2021). Feasibility study on microwave welding of thermoplastic using multiwalled carbon nanotubes as susceptor. *Nanomaterials and Nanotechnology*, 11, 18479804211002926.

Gupta, D., Singh, S., Jain, V., & Kumar, R. (2015). Joining of bulk cast iron through microwave energy. *International Journal for Technological Research in Engineering*, 2(7), 1079–1084.

Hong, D., Yuan, J., Yin, Z., Peng, H., & Zhu, Z. (2020). Ultrasonic-assisted preparation of complex-shaped ceramic cutting tools by microwave sintering. *Ceramics International*, 46(12), 20183–20190.

James, S. J. (2005). Refrigeration and the safety of poultry meat. In S. J. James (Ed.), *Food Safety Control in the Poultry Industry* (p. 333). University of Bristol, UK.

Kaushal, S. (2022). Microstructure and tribological characterization of composite castings developed through in-situ microwave hybrid heating. *International Journal of Metalcasting*, 16(4), 076405.

Kaushal, S., Gupta, D., & Bhowmick, H. (2018). An approach for functionally graded cladding of composite material on austenitic stainless steel substrate through microwave heating. *Journal of Composite Materials*, 52(3), 301–312.

Kaushal, S., Gupta, D. and Bhowmick, (2021). Wear behavior of microwave processed Ni-WC8Co-based functionally graded materials. *Proceedings of the Institution of Mechanical Engineers, Part L: Journal of Materials: Design and Applications*, 235(5), 1036–1045.

Kaushal, S., Kumari, S., Mudgal, D., Gupta, D., & Vasudev, H. (2023). Experimental studies on the surface characteristics of bimetallic joints interface fabricated through microwave irradiation. *Surface Review and Letters*, 30(10), 1–14.

Kaushal, S., Singh, D., Gupta, D., & Jain, V. (2019). Processing of Ni-WC-Cr_3C_2-based metal matrix composite cladding on SS-316L substrate through microwave irradiation. *Journal of Composite Materials*, 53(8), 1023–1032.

Kazemi, F., Boehm, G., & Arnold, T. (2019). Development of a model for ultra-precise surface machining of N-BK7® using microwave-driven reactive plasma jet machining. *Plasma Processes and Polymers*, 16(12), 1900119.

Ku, H. S., Siores, E., & Ball, J. A. (1999). Microwave facilities for welding thermoplastic composites and preliminary results. *Journal of Microwave Power and Electromagnetic Energy*, 34(4), 195–205.

Kumar, A., Jain, V. G., & Gupta, D. G. (2015). Microwave Joining of Stainless Steel and Their Characterizations (Doctoral dissertation).

Kumar, A., Sehgal, S., Singh, S., & Bagha, A. K. (2020). Joining of SS304-SS316 through novel microwave hybrid heating technique without filler material. *Materials Today: Proceedings*, 26, 2502–2505.

Kumar, G., & Sharma, A. K. (2018). Role of dielectric fluid and concentrator material in microwave drilling of borosilicate glass. *Journal of Manufacturing Processes*, 33, 184–193.

Kumar, R., Bhowmick, H., Gupta, D., & Bansal, S. (2021). Development and characterization of multiwalled carbon nanotube-reinforced microwave sintered hybrid aluminum metal matrix composites: An experimental investigation on mechanical and tribological performances. *Proceedings of the Institution of Mechanical Engineers, Part L: Journal of Materials: Design and Applications*, 235(10), 2310–2323.

Leo, P., D'Ostuni, S., & Casalino, G. (2016). Hybrid welding of AA5754 annealed alloy: Role of post weld heat treatment on microstructure and mechanical properties. *Materials & Design*, 90, 777–786.

Lingappa, M. S., Srinath, M. S., & Amarendra, H. J. (2017). Microstructural and mechanical investigation of aluminium alloy (Al 1050) melted by microwave hybrid heating. *Materials Research Express*, 4(7), 076504.

Mishra, R. R., & Sharma, A. K. (2019). Multi-physics simulation of in situ microwave casting of 7039 Al alloy inside different applicators and cast microstructure. *Proceedings of the Institution of Mechanical Engineers, Part E: Journal of Process Mechanical Engineering*, 233(3), 617–629.

Mishra, S., Sahoo, S. S., Debnath, A. K., Muthe, K. P., Das, N., & Parhi, P. (2020). Cobalt ferrite nanoparticles prepared by microwave hydrothermal synthesis and adsorption efficiency for organic dyes: Isotherms, thermodynamics and kinetic studies. *Advanced Powder Technology*, 31(11), 4552–4562.

Oda, S. J. (1992). Microwave remediation of hazardous waste: A review. *Microwave Processing of Materials III*, 269, 453–464.

Pal, M., Sehgal, S., & Kumar, H. (2020). Optimization of elemental weight% in microwave-processed joints of SS304/SS316 using Taguchi philosophy. *Journal of Advanced Manufacturing Systems*, 19(03), 543–565.

Pal, M., Sehgal, S., Kumar, H., & Singh, A. P. (2018). Manufacturing of joints of stainless steels through microwave hybrid heating. *Materials Today: Proceedings*, 5(14), 28149–28154.

Pan, J. P., Zhu, X. W., & Tan, L. J. (2013). Detecting cold weld defects in HDPE piping thermal fusion welds based on microwave technique. *Applied Mechanics and Materials*, 423, 852–856.

Perumalla, S. K., Muniyappa, A., & Sivanandam, A. (2020). Applications of microwave heat treatment process to enhance the surface properties of bronze and steel materials used in journal bearing. *Proceedings of the Institution of Mechanical Engineers, Part C: Journal of Mechanical Engineering Science*, 234(10), 2064–2076.

Potente, H., Karger, O., & Fiegler, G. (2003). Heatability of plastics in the microwave field—investigations on direct and indirect mw-welding. *Welding in the World*, 47, 25–30.

Rosa, R., Veronesi, P., Han, S., Casalegno, V., Salvo, M., Colombini, E., Leonelli, C., & Ferraris, M. (2013). Microwave assisted combustion synthesis in the system Ti–Si–C for the joining of SiC: Experimental and numerical simulation results. *Journal of the European Ceramic Society*, 33(10), 1707–1719.

Rosenberg, U., & Bogl, W. (1987). Microwave pasteurization, sterilization, blanching, and pest control in the food industry. *Food Technology* (USA), 41(6).

Sahota, D. S., Bansal, A., & Kumar, V. (2021). Application of microwave in welding of metallic materials–A review. *Materials Today: Proceedings*, 43, 466–470.

Saloni, Singh, G., Dinesh, and Kaushal, S., (2022). Microwave processed EWAC/SiC based metal matrix composite castings, *Materials Today Proceedings*, 50(5), 842–847.

Samyal, R., Bagha, A. K., & Bedi, R. (2020a). Microwave joining of similar/dissimilar metals and its characterizations: A review. *Materials Today: Proceedings*, 26, 423–433.

Samyal, R., Bagha, A. K., & Bedi, R. (2020b). The casting of materials using microwave energy: A review. *Materials Today: Proceedings*, 26, 1279–1283.

Savu, S. V., Savu, I. D., & Benga, G. C. (2016). Heat affected zone in microwave polymer welding. *Advanced Materials Research*, 1138, 165–171.

Schiffmann, R. F. (2001). Microwave processes for the food industry. In *Handbook of Microwave Technology for Food Application* (pp. 299–338). CRC Press.

Shukla, M., Ghosh, S., Dandapat, N., Mandal, A. K., & Balla, V. K. (2016). Microwave-assisted brazing of alumina ceramics for electron tube applications. *Bulletin of Materials Science*, 39, 587–591.

Silberglitt, R., Ahmad, I., Black, W. M., & Katz, J. D. (1993). Recent developments in microwave joining. *MRS Bulletin*, 18(11), 47–50.

Singh, B., Kaushal, S., Gupta, D., and Bhowmick, H. (2018). On development and dry sliding wear behavior of microwave processed Ni/Al_2O_3 composite clad. *Journal of Tribology*, 140(6), 061603.

Singh, S., Gupta, D., & Jain, V. (2016). Recent applications of microwaves in materials joining and surface coatings. *Proceedings of the Institution of Mechanical Engineers, Part B: Journal of Engineering Manufacture*, 230(4), 603–617.

Singh, S., Gupta, D., Jain, V., & Sharma, A. K. (2015). Microwave processing of materials and applications in manufacturing industries: a review. *Materials and Manufacturing Processes*, 30(1), 1–29.

Singh, S., Singh, P., Gupta, D., Jain, V., Kumar, R., & Kaushal, S. (2019). Development and characterization of microwave processed cast iron joint. *Engineering Science and Technology, an International Journal*, 22(2), 569–577.

Singh, S., Suri, N. M., & Belokar, R. M. (2015). Characterization of joint developed by fusion of aluminum metal powder through microwave hybrid heating. *Materials Today: Proceedings*, 2(4–5), 1340–1346.

Siores, E., & Do Rego, D. (1995). Microwave applications in materials joining. *Journal of Materials Processing Technology*, 48(1–4), 619–625.

Soni, P., Sehgal, S., Kumar, H., & Singh, A. P. (2018). Joining of SS316-SS316 through microwave hybrid heating by using Nickel nano-powder. *International Journal of Applied Engineering Research*, 8(8), 6446–6449.

Sosa, E. D., Worthy, E. S., & Darlington, T. K. (2016). Microwave assisted manufacturing and repair of carbon reinforced nanocomposites. *Journal of Composites*, 16, Article ID 7058649.

Srinath, M. S., Murthy, P. S., Sharma, A. K., Kumar, P., & Kartikeyan, M. V. (2012). Simulation and analysis of microwave heating while joining bulk copper. *International Journal of Engineering, Science and Technology*, 4(2), 152–158.

Srinath, M. S., Sharma, A. K., & Kumar, P. (2011). Investigation on microstructural and mechanical properties of microwave processed dissimilar joints. *Journal of Manufacturing Processes*, 13(2), 141–146.

Sun, X., Wu, G., Yu, J., & Du, C. (2018). Efficient microwave welding of polypropylene using graphite coating as primers. *Materials Letters*, 220, 245–248.

Sun, X., Yu, J., & Wu, G. (2019). Study on microwave welding of polypropylene by carbon nanotube. *Integrated Ferroelectrics*, 197(1), 16–22.

Tamang, S., & Aravindan, S. (2021). Effect of susceptors on joining of AA6061-T6 to AZ31B by microwave hybrid heating. *Journal of Materials Engineering and Performance*, 31, 1–10.

Varadan, V. K., & Varadan, V. V. (1991). Microwave joining and repair of composite materials. *Polymer Engineering & Science*, 31(7), 470–486.

Varghese, K. S., Pandey, M. C., Radhakrishna, K., & Bawa, A. S. (2014). Technology, applications and modelling of ohmic heating: A review. *Journal of Food Science and Technology, 51*, 2304–2317.

Wise, R. J., & Froment, I. D. (2001). Microwave welding of thermoplastics. *Journal of Materials Science*, 36, 5935–5954.

Wu, C. Y., & Benatar, A. (1997). Microwave welding of high density polyethylene using intrinsically conductive polyaniline. *Polymer Engineering & Science*, 37(4), 738–743.

Yarlagadda, P. K., & Chai, T. C. (1998). An investigation into welding of engineering thermoplastics using focused microwave energy. *Journal of Materials Processing Technology*, 74(1–3), 199–212.

3 Microwave Cladding for Enhanced Tribological Performance

Bhupinder Singh

3.1 INTRODUCTION

The multidisciplinary field of surface engineering includes numerous methods and strategies to alter the surface characteristics of materials. It entails adding new surface functionalities, changing surface structures, or applying specialized coatings to improve the functionality, performance, and durability of materials in a variety of applications. Surface engineering strives to optimize the surface's thermal, electrical, optical, and chemical properties as well as its resistance to wear, corrosion, friction, and fatigue.

Surface engineering techniques include a wide range of procedures, including thin films, thermal spray coatings, and chemical vapour deposition. Surface treatments such as ion implantation, laser surface modification, plasma nitriding, and electroplating are also used. To obtain desired surface qualities, surface engineering may also include surface texturing, nanoscale patterning, and the insertion of functional additives.

On a global scale, various sectors such as transport, industry, domestic, agriculture, and forestry consumed a total of 396 Exajoules (EJ) of primary energy in 2014 [1]. Studies have shown that approximately 23% of energy is lost due to wear and frictional rubbing [2]. Friction and wear are connected occurrences that happen when two moving surfaces come into contact, such as in bearings, cylinder liners, and cam followers. A research reported by Holmberg et al. revealed that by leveraging advanced tribological technologies, wear and friction losses can be lowered by around 18–40% [3]. The greatest potential for energy savings lies in the power generation and transportation sectors.

Throughout history, the development of new coatings has been the favoured approach in the field of tribology for reducing friction and wear and achieving energy savings. Among these techniques, thermal spray has gained significant popularity in industrial applications. However, it still has certain limitations, such as being mechanically bonded with high surface roughness [4]. Additionally, from an economic standpoint, conventional heating methods for coating materials result in high energy

DOI: 10.1201/9781003449225-3

losses. Consequently, the quality of the coating can be affected by the actions involved in the deposition process, while the use of appropriate energy sources can help mitigate environmental and economic aspects.

Therefore, a new research approach has appeared, which focuses on utilizing nonconventional energy sources like solar power, lasers, and microwaves to engineer surfaces with improved tribological properties. By exploring these alternative energy sources, researchers aim to overcome the limitations of conventional techniques, enhance coating quality, and minimize energy consumption and environmental impact. This approach holds great potential for advancing the field of tribology and achieving substantial energy savings in various industrial sectors.

3.2 DIFFERENT TECHNIQUES IN SURFACE COATING/CLADDINGS

To improve the performance and durability of materials, surface coating techniques are frequently applied. Thermal spray, laser cladding, and weld overlays are three commonly used methods in surface engineering. Each method has its own principles and makes a distinctive contribution to the area.

Thermal Spray: In thermal spray, a coating material is deposited onto a substrate using a thermal energy source, which is a flexible and widely accepted surface engineering approach. A gas or plasma stream is used to drive the coating material onto the substrate. The material is melted or heated during the pathway in a high-temperature setting. The coating material sticks to the substrate and forms a coating as shown in Figure 3.1 [5]. A wide variety of materials, including metals, ceramics, polymers, and composites, can be used using this approach. Improvements in corrosion resistance, wear resistance, thermal insulation, electrical conductivity, and aesthetic appeal are just a few advantages that thermal spray coatings can offer [6].

Laser cladding: Material is deposited onto a substrate using a high-energy laser beam in the precise and localized surface engineering process known as laser cladding as shown in Figure 3.2. The coating material is melted by the laser beam and then fused to the substrate to form a metallurgical bond [7]. A wide variety of materials, including metals, alloys, and ceramics, can be deposited using laser cladding because it provides exceptional control over the heat input and dilution [8]. This method is frequently used to fix and restore broken parts, increase wear resistance, and add certain qualities like corrosion resistance or high-temperature resistance.

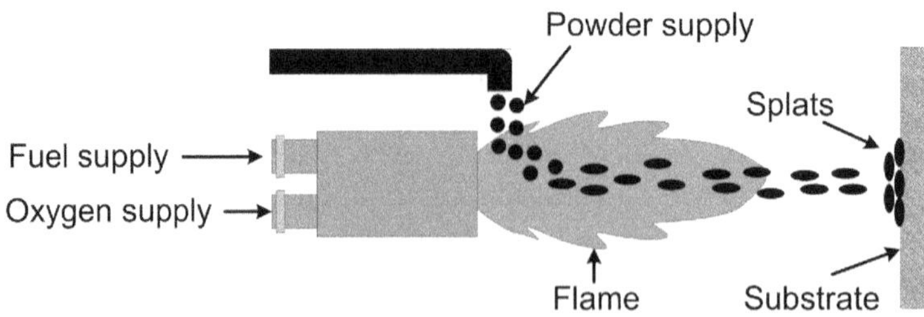

FIGURE 3.1 Principle of thermal spraying technology.

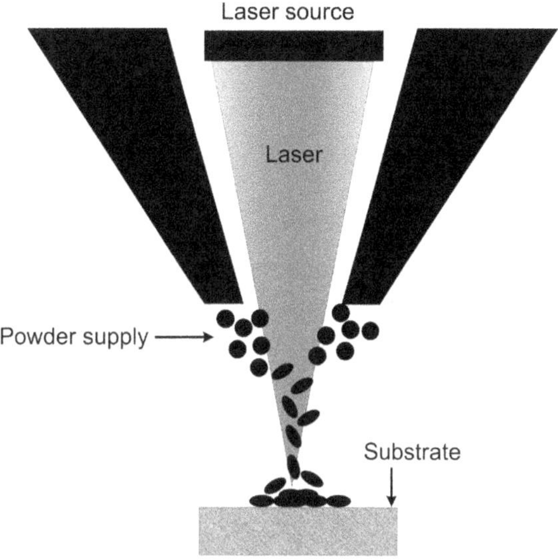

FIGURE 3.2 Principle of laser cladding.

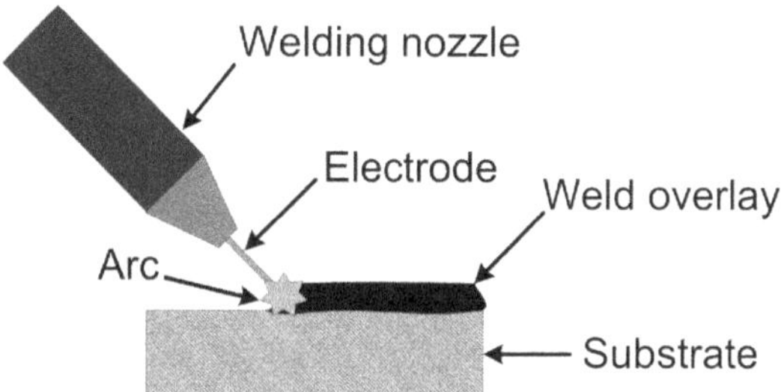

FIGURE 3.3 Principle of weld overlays.

Weld overlays: Another surface engineering method, weld overlays, entails welding a layer of material onto a substrate. A consumable electrode or filler material is often melted throughout the procedure and fused with the base metal to form a metallurgical bond as shown in Figure 3.3 [9]. Different welding techniques, such as shielded metal arc welding, gas tungsten arc welding, or gas metal arc welding, can be used to apply weld overlays. They are frequently utilized to improve components' surface qualities, such as wear resistance, corrosion resistance, and high-temperature resistance. Weld overlays can also be used to restore or repair damaged components. In conclusion, surface engineering techniques like thermal spray, laser cladding, and weld overlays are significant and make unique contributions. Various materials can be deposited onto substrates via thermal spray, giving them improved qualities. With the help of laser cladding, surfaces can be modified precisely and locally, simplifying repairs and the addition of traits. Weld overlays use welding techniques to enhance surface qualities. These methods are crucial to the production of materials with greater performance and longer lives in sectors like industrial, aerospace, automotive, and energy.

3.3 MICROWAVE CLADDING

3.3.1 Introduction to Microwaves

Microwaves, which fall on the electromagnetic spectrum between radio waves and infrared radiation, have unique characteristics that distinguish them from other types of electromagnetic radiation. Microwaves are valuable in a wide range of applications due to their efficiency in penetrating objects, interacting with matter, and propagating through the atmosphere.

The fundamental principles of microwaves revolve around their frequency range, typically spanning from 300 MHz to 300 GHz, corresponding to wavelengths ranging from one millimetre to one metre [10]. Microwaves exhibit unique behaviours within this range, such as absorption, reflection, and transmission which are influenced by the properties of the medium. Understanding how microwaves propagate through waveguides, resonant structures, and transmission lines is crucial for harnessing their potential in various applications.

3.3.2 Microwave–Material Interaction

Electromagnetic radiation interacts with the ions, molecules, and electrons present in a target material to generate microwave heating. In contrast to traditional heating techniques, this approach has several unique qualities. First, microwave heating uses an "inside-out" heating pattern in which the material itself produces the heat [11]. This is in contrast to the conventional techniques that rely on heat transfer from outside sources.

Second, the temperature variation within the material is also relatively minor since microwave heating is linked with a low thermal gradient. This even heating guarantees that the material is heated evenly throughout, lowering the possibility of localized overheating or thermal damage.

Third, the selective nature of microwave heating is another noteworthy quality. The electromagnetic radiation can be adjusted to interact only with the target substance rather than the surrounding materials. By limiting the heating of nearby materials, this selectivity enables the targeted heating of selected components. Thus, microwave heating provides more control and precision over the heating process.

Finally, the energy efficiency of microwave heating is an unmatchable benefit. Microwave heating uses less energy because the heat is internally created rather than relying just on heat transfer from the outside. This decrease in energy demand may result in cost savings and increased general heating application efficiency.

3.3.3 Principle of Microwave Cladding

The Maxwell–Wagner Polarization effect is a phenomenon that causes electrical losses inside the material. It is essential that the sample size and the microwave penetration depth be roughly similar in order to accomplish efficient heating. Therefore, both material size and penetration depth should be taken into account to determine the ability of the material to transform microwave radiation into heat.

However, microwave processing is substantially more difficult when dealing with metallic materials. This is because, under normal circumstances, metallic materials reflect microwaves with a frequency of 2.45 GHz. The reason for this reflection is that metallic materials have a very shallow penetration depth or skin depth. As a result, the metallic material is difficult to heat up since the microwave radiation is unable to efficiently penetrate it. The skin depth of a given material can be expressed as in Equation (3.1) [12]:

$$\delta = \sqrt{\frac{\rho_e}{\pi f \mu_r \mu_0}} \tag{3.1}$$

where, δ is skin depth (μm), ρ_e is the resistivity (μΩ-cm), f is the frequency (GHz), μ_r is the relative permeability, and μ_o is the absolute permeability (H/m). The depth to which microwaves can enter metal can be impacted by temperature-dependent features like magnetic permeability and resistivity during the microwave heating process. A method known as microwave hybrid heating, which uses a lossy susceptor to absorb microwave radiation and transmit thermal energy to powder particles via conduction and convection, has been developed to overcome the challenge with metal processing as shown in Figure 3.4. Once the clad powder reaches a critical temperature, direct microwave absorption takes place, producing rapid heat generation and thus melting. The proximity to the hot melt pool also causes the thin substrate layer to melt, enabling elemental diffusion along the interface and the creation of a metallurgical link between the clad and the substrate.

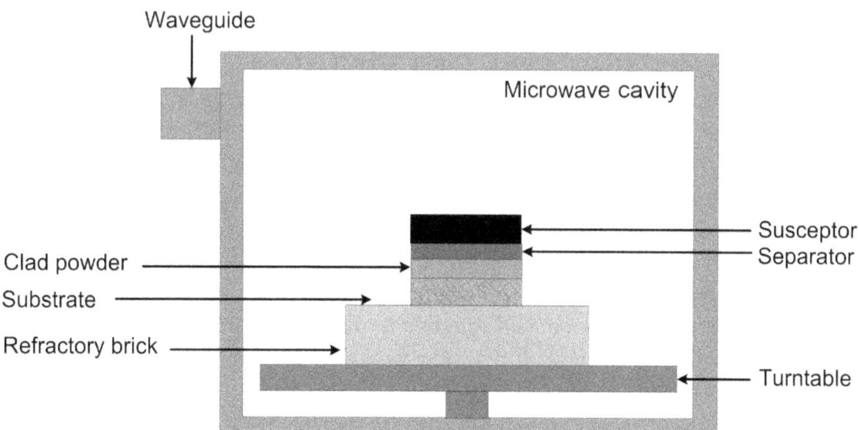

FIGURE 3.4 Principle of microwave cladding.

Microwave cladding offers many benefits over conventional surface alteration techniques, including the development of a consistent microstructure, low porosity, and less clad dilution, leading to superior tribological performance [13–19].
Zafar and Sharma developed a WC-12Co clad on AISI-304 utilizing the microwave cladding [20]. The resulting clad had a thickness of 1 mm and had a distinctive structure with skeleton-like carbides embedded inside the metallic matrix. The mutual diffusion of elements allowed a metallurgical bonding between the clad and the substrate. The clad's average hardness was found to be 1135 HV. A pyrometer was used to establish the ideal temperature for clad formation, and the results showed that 1400°C is the minimum temperature for WC-12Co clad formation.

Further, Zafar and Sharma [21] compared the abrasive and erosive wear behaviour of nanometric and micrometric clads. The presence of nano-carbides in the nanometric clad resulted in a decreased free mean path, which in turn increased the microhardness of the clad. The researchers also found that when compared to the micrometric clad, the nanometric clad showed a 1.6-fold reduction in abrasive wear and a 3-fold reduction in erosive wear.

Singh et al. suggested using composite clads to improve wear performance [22]. It was reported that Ni/Al_2O_3 composite clad when compared to the SS-304 substrate displayed a remarkable 156 times greater wear resistance. This improved wear performance was linked to characteristics of the composite clad such as a consistent microstructure, the lack of pores, and microcracks.

Overall, improvements in tribological performance, homogenous microstructures, and less wear have all been made possible by advances in microwave cladding.

3.3.4 Temperature Characteristics in Microwave Cladding

Microwave cladding is gaining popularity as a surface modification technology due to its inherent potential to change the surface characteristics of materials. However, the capacity to precisely regulate the temperature throughout the procedure is a crucial component of any surface modification technology. However, the measurement of temperature inside the electromagnetic field is another challenge. Different temperature monitoring techniques have been used; however, they frequently only offer a limited amount of information. Gangurde et al. reported the use of optical fibres placed inside the microwave cavity to measure temperature without disrupting the electromagnetic field. Unfortunately, optical fibres can only be used successfully in processes that are carried out at temperatures below 300°C [23].

Another strategy used by Will et al. entailed swiftly placing a thermocouple into a catalytic bed to measure the temperature after the microwave oven was turned off [24]. Even though this technology made it possible to gauge temperature, it did not offer ongoing real-time monitoring. As a result, methods for measuring temperature without any contact with the material indirectly have become popular in microwave processing. Nigar et al. validated their modelling results by using a thermographic camera to record the temperature distribution of microwave-heated zeolite [25]. Ge et al. reported on infrared temperature measurements while treating Chinese coal

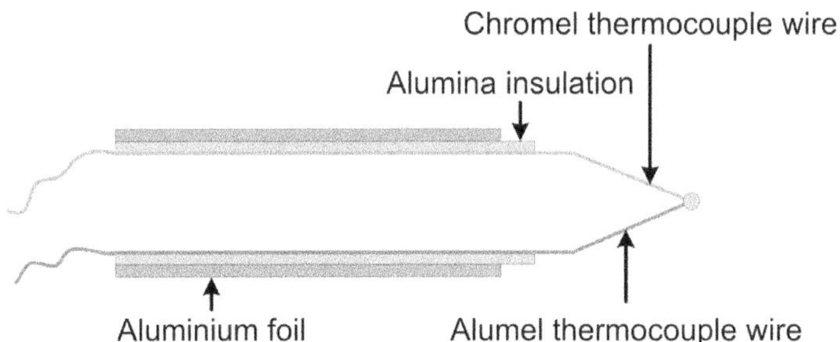

FIGURE 3.5 Schematic representation of modified metal thermocouple.

with a microwave [26]. Undri et al. measured the temperature of the reactor wall during the microwave pyrolysis of waste tyres using infrared thermometry [27].

In the case of microwave cladding, some researchers have used high-temperature range infrared (IR) pyrometers to track the temperature inside the microwave cavity. It is crucial to keep in mind that IR pyrometers only give information on the temperature distribution of the susceptor material and not the clad material. Due to the selective heating patterns associated with microwaves, this constraint hinders the management of the cladding process since knowing merely the susceptor temperature is insufficient. Modified metal thermocouples have been used to monitor the temperature inside the electromagnetic field. A modified B-type thermocouple with a metal sheath made of aluminium to reflect microwaves and prevent interaction with thermocouple wires is shown in Figure 3.5. However, an electromagnetic field can cause a point discharge phenomenon with arcing at the thermocouple tip. To avoid sparking during microwave processing, a thermocouple was grounded with wire prior to microwave exposure.

The potential goal of employing a modified B-type thermocouple is to evaluate the clad powder's real-time temperature characteristics. Continuous temperature measurements are possible with this modified thermocouple during microwave-cladding Ni-based powder. The modified thermocouple provides information on the temperature of the clad material itself, in contrast to the IR pyrometer, which measures the susceptor's surface temperature. For establishing the critical temperature of the clad powder and streamlining the microwave cladding procedure, this real-time temperature distribution data is essential.

The critical temperature of the Ni powder during microwave cladding was found to be approximately 454°C using a modified metal thermocouple. Researchers and engineers can achieve desired results and guarantee the quality of the clad material by adjusting process parameters, such as microwave power and time, with the help of this information.

3.3.5 Numerical Simulation of Microwave Cladding Process

The distribution of temperature across the clad layer is difficult to measure experimentally due to the complex nature of the interaction between microwaves and materials within the electromagnetic field. Therefore, using numerical simulation techniques can greatly reduce time and resource requirements while providing a more thorough knowledge of the complex physical processes involved in microwave cladding.

Researchers reported theoretical tools to build numerical models that depict the microwave-induced heating of materials by utilizing Maxwell's equations, which offer a mathematical framework for describing electromagnetic fields, and the heat transfer equation, which characterizes temperature dynamics as shown in Figure 3.6.

It realistically recreates and investigates the complicated interactions between microwaves and materials, which sheds light on the complex physical events that take place throughout the cladding process. These simulations help to optimize the cladding process by properly simulating the electromagnetic fields and temperature distribution.

Furthermore, the use of numerical simulations eliminates the necessity for expensive and time-consuming experimental setups. Researchers can instead iteratively modify their models, experimenting with different scenarios and settings to acquire a better understanding of the microwave–material interaction. This method enables researchers to examine the impacts of various microwave settings, material qualities, and geometries in a timely and cost-effective manner.

In conclusion, it is difficult to determine the temperature distribution during microwave cladding experimentally because of the intricacy of the interactions between microwaves and materials in the electromagnetic field. However, researchers may create theoretical models that provide helpful insights into the complex physical processes at work by using numerical simulations based on Maxwell's equations and the heat transfer equation. In comparison to solely experimental approaches, these models offer a thorough knowledge of microwave-induced heating and facilitate the optimization of the cladding process.

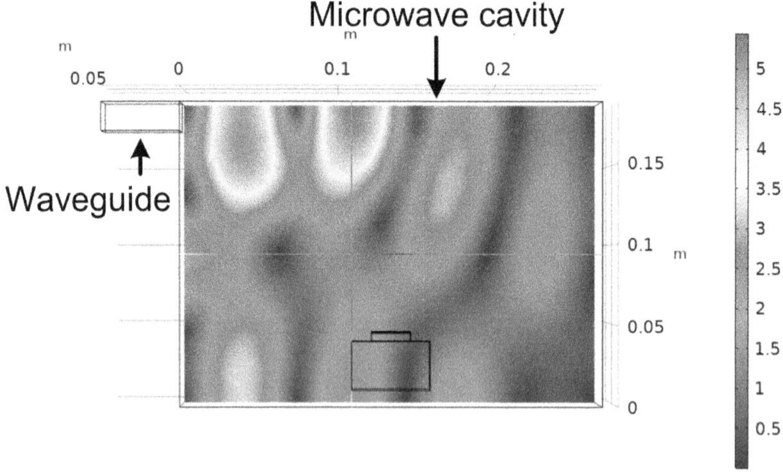

FIGURE 3.6 Numerical simulation of the distribution of electric field intensity in the microwave cladding process.

3.4 CONCLUSION

Microwave cladding has shown its effectiveness in achieving desirable surface modifications for better tribological performance. Several studies have shown that various materials can be successfully clad, resulting in improved wear resistance and microstructures. Using temperature monitoring systems, such as modified thermocouples, researchers were able to optimize the cladding process by precisely measuring and controlling the temperature. Microwave cladding has the potential to significantly advance the field of surface engineering by providing superior microstructures, low porosity, and increased wear resistance in a variety of industrial sectors.

3.5 LIMITATIONS

Microwave cladding is limited to simpler geometry. To clad large components, a larger microwave applicator with an expanded cavity will be required. Inadequate experimental data hinders the simulation of the microwave cladding process, limiting its flexibility. The lack of critical experimental data, specifically temperature-dependent physical and dielectric properties, makes accurate temperature prediction difficult.

REFERENCES

[1] International Energy Agency, *Key World Energy Statistics* (2016), n.d.

[2] K. Holmberg, A. Erdemir, Influence of tribology on global energy consumption, costs and emissions, *Friction.* 5 (2017) 263–284. https://doi.org/10.1007/s40544-017-0183-5.

[3] K. Holmberg, A. Erdemir, The impact of tribology on energy use and CO_2 emission globally and in combustion engine and electric cars, *Tribol Int.* 135 (2019) 389–396. https://doi.org/10.1016/j.triboint.2019.03.024.

[4] G. Prashar, H. Vasudev, L. Thakur, Performance of different coating materials against slurry erosion failure in hydrodynamic turbines: A review, *Eng Fail Anal.* 115 (2020). https://doi.org/10.1016/j.engfailanal.2020.104622.

[5] H.S. Grewal, A. Agrawal, H. Singh, Slurry erosion performance of Ni-Al_2O_3 based composite coatings, *Tribol Int.* 66 (2013) 296–306. https://doi.org/10.1016/j.triboint.2013.06.010.

[6] A.S. Khanna, S. Kumari, S. Kanungo, A. Gasser, Hard coatings based on thermal spray and laser cladding, *Int J Refract Metals Hard Mater.* 27 (2009) 485–491. https://doi.org/10.1016/j.ijrmhm.2008.09.017.

[7] Y. Liang, Z.Y. Liao, L.L. Zhang, M.W. Cai, X.S. Wei, J. Shen, A review on coatings deposited by extreme high-speed laser cladding: processes, materials, and properties, *Opt Laser Technol.* 164 (2023). https://doi.org/10.1016/j.optlastec.2023.109472.

[8] M. E, H.X. Hu, X.M. Guo, Y.G. Zheng, Comparison of the cavitation erosion and slurry erosion behavior of cobalt-based and nickel-based coatings, *Wear.* 428–429 (2019) 246–257. https://doi.org/10.1016/j.wear.2019.03.022.

[9] S.A. Romo, J.F. Santa, J.E. Giraldo, A. Toro, Cavitation and high-velocity slurry erosion resistance of welded Stellite 6 alloy, *Tribol Int.* 47 (2012) 16–24. https://doi.org/10.1016/j.triboint.2011.10.003.

[10] S. Tabar Maleki, M. Babamoradi, Microwave absorption theory and recent advances in microwave absorbers by polymer-based nanocomposites (carbons, oxides, sulfides, metals, and alloys), *Inorg Chem Commun.* 149 (2023). https://doi.org/10.1016/j.inoche.2023.110407.

[11] W. Tayier, S. Janasekaran, V.C. Tai, Microwave hybrid heating (MHH) of Ni-based alloy powder on Ni and steel-based metals – a review on fundamentals and parameters, *International Journal of Lightweight Materials and Manufacture.* 5 (2022) 58–73. https://doi.org/10.1016/j.ijlmm.2021.10.002.

[12] D. Gupta, A.K. Sharma, Microwave cladding: A new approach in surface engineering, *J Manuf Process.* 16 (2014) 176–182. https://doi.org/10.1016/j.jmapro.2014.01.001.

[13] S. Kaushal, D. Gupta, and H. Bhowmick, On microstructure and wear behavior of microwave processed composite clad, *J Tribol.* 139 (2017) 061602. doi: 10.1115/1.4035844.

[14] S. Kaushal, B. Singh, D. Gupta, H. Bhowmick, and V. Jain, An approach for developing nickel–alumina powder-based metal matrix composite cladding on SS-304 substrate through microwave heating, *J Compos Mater.* 52(2018) 2131–2138. doi: 10.1177/0021998317740732.

[15] S. Kaushal, D. Gupta, and H. Bhowmick, Investigation of dry sliding wear behavior of Ni–SiC microwave cladding, *J Tribol.* 139(2017) 41603. https://doi.org/10.1115/1.4035147

[16] Saloni, G. Singh, Dinesh, and S. Kaushal, Microwave processed EWAC/SiC based metal matrix composite castings, *Mater Today Proc*, 50 (2022) 842–847. https://doi.org/10.1016/j.matpr.2021.06.064

[17] S. Bansal *et al.*, Effect of variation of WC reinforcement on metallurgical and cavitation erosion behavior of microwave processed NiCrSiC-WC composites clads, *Proc. Inst. Mech Eng Part C J Mech Eng Sci.* 237(2023) 5460–5475, https://doi.org/10.1177/09544062231164517

[18] S. Kaushal, D. Singh, D. Gupta, and V. Jain, Processing of Ni–WC–Cr_3C_2-based metal matrix composite cladding on SS-316L substrate through microwave irradiation, *J Compos Mater.* 53(2019) 1023–1032. https://doi.org/10.1177/0021998318794846

[19] S. Kaushal and S. Singh, Processing strategy for high strength Ni-based hybrid composite clad on SS 316L steel through microwave heating, *Proc. Inst. Mech. Eng., Part B.* 236(2022) 190–203, https://doi.org/10.1177/09544054211021360

[20] S. Zafar and A.K. Sharma, Development and characterisations of WC-12Co microwave clad, *Mater Charact.* 96 (2014) 241–248. https://doi.org/10.1016/j.matchar.2014.08.015.

[21] S. Zafar and A.K. Sharma, Abrasive and erosive wear behaviour of nanometric WC-12Co microwave clads, *Wear.* 346–347 (2016) 29–45. https://doi.org/10.1016/j.wear.2015.11.003.

[22] B. Singh, S. Kaushal, D. Gupta, and H. Bhowmick, On development and dry sliding wear behavior of microwave processed Ni/ Al_2O_3 composite clad, *J Tribol.* 140 (2018). https://doi.org/10.1115/1.4039996.

[23] L.S. Gangurde, G.S.J. Sturm, T.J. Devadiga, A.I. Stankiewicz, and G.D. Stefanidis, Complexity and challenges in noncontact high temperature measurements in microwave-assisted catalytic reactors, *Ind Eng Chem Res.* 56 (2017) 13379–13391. https://doi.org/10.1021/acs.iecr.7b02091.

[24] H. Will, P. Scholz, and B. Ondruschka, Heterogeneous gas-phase catalysis under microwave irradiation-a new multi-mode microwave applicator, n.d.

[25] H. Nigar, G.S.J. Sturm, B. Garcia-Baños, F.L. Peñaranda-Foix, J.M. Catalá-Civera, R. Mallada, A. Stankiewicz, and J. Santamaría, Numerical analysis of microwave heating

cavity: Combining electromagnetic energy, heat transfer and fluid dynamics for a NaY zeolite fixed-bed, *Appl Therm Eng.* 155 (2019) 226–238. https://doi.org/10.1016/j.applthermaleng.2019.03.117.

[26] L. Ge, Y. Zhang, Z. Wang, J. Zhou, and K. Cen, Effects of microwave irradiation treatment on physicochemical characteristics of Chinese low-rank coals, *Energy Convers Manag.* 71 (2013) 84–91. https://doi.org/10.1016/j.enconman.2013.03.021.

[27] A. Undri, S. Meini, L. Rosi, M. Frediani, and P. Frediani, Microwave pyrolysis of polymeric materials: Waste tires treatment and characterization of the value-added products, *J Anal Appl Pyrolysis, Elsevier B.V.* 2013, 149–158. https://doi.org/10.1016/j.jaap.2012.11.011.

4 Microwave-Processed Casting

A Green Manufacturing Approach

Jatinder Pal, Dheeraj Gupta, and Tejinder Paul Singh

4.1 INTRODUCTION

4.1.1 BACKGROUND

The manufacturing sector dramatically impacts the environment and the world's energy use (Al-Shetwi, 2022). Commonly employed in many industries, conventional casting techniques frequently involve high energy costs, significant material waste, and toxic pollutants (Pal, Gupta, & Singh, 2022). Innovative solutions that can lessen the environmental impact of casting processes are in high demand as the requirement for environmentally responsible and sustainable industrial practices increases (Kushwaha & Sharma, 2016; Singh, Gupta, & Jain, 2016). This study aims to investigate how microwave processing can be used in the casting industry as a green production strategy. Based on electromagnetic energy, microwave processing can eliminate waste, improve material qualities, speed up processing times, and use less energy. We can open the door to more environmentally friendly production procedures by solidly understanding the principles, benefits, and difficulties of microwave-treated casting.

The casting manufacturing technique has been extensively employed to create various types of parts used in numerous areas for more than 5000 years (Jones & Yuan, 2003). Casting a significant volume of material costs around one-tenth as much to produce as alternative competitive technologies (Hashim, Looney, & Hashmi, 1999). The earlier casting method was utilized to make essential tools, further followed by the development of jewelry and creative items from bronze and gold (Singh, Singh, & Hashmi, 2016). But at the start of World War II, the industrial sector realized that there was a need for intricate parts and adequate materials for aerospace, defense, automobile, and many other industries. After World War II, these parts were produced using a variety of conventional and some advanced casting techniques such as green sand casting, gravity die casting, pressure die

 DOI: 10.1201/9781003449225-4

casting, molding, centrifugal casting, lost-foam casting, vacuum casting, squeezing casting, continuous casting, shell molding, and so on, but these methods have their drawbacks in the form of defects such as hot tears, vermicular, lack in fluidity, cold shuts, mismatches, crack initiation, blowholes, pores, unfilled cavity, fettling, shrinkage, and difficulty to process thin parts.

Additionally, traditional casting manufacturing methods have certain drawbacks: higher costs, more energy usage, lengthy processing times, and pollution. To keep in mind the limits of current typical casting processes, researchers investigated cutting-edge nonconventional manufacturing technologies such as 3D digital printing, inkjet printing, fused deposition modeling, laminated object-oriented manufacturing, selective laser melting (SLM), sintering, powder metallurgy route, stir-squeeze casting (Buchbinder, Schleifenbaum, Heidrich, Meiners, & Bültmann, 2011; Sama, Wang, & Manogharan, 2018; Silvestroni & Sciti, 2010). Still, these techniques have their limits in the form of high initial and running costs, and high skill and strong technical knowledge are required.

Researchers also created innovative, cost-effective, and environmentally friendly processes that may provide clean and green energy at the lowest possible cost due to their desire for environmental safety (Kushwaha & Sharma, 2016). Microwave energy for melting and casting metallic materials is one of the most significant methods for green and environment-friendly manufacturing. Microwave energy in the form of microwave hybrid heating (MHH) has been used to get around the challenges that the manufacturing mentioned above (conventional and nonconventional) processes confront (Lingappa, Srinath, & Amarendra, 2018; Pal et al., 2022; Singh et al., 2016; Singh, Gupta, & Jain, 2018). The outside to inside and inside to outside, and homogeneous volumetric heating with fewer thermal gradients, pollution-free, lessened processing time, and fewer defects are the significant essential features of MHH. Additionally, it improves the casting's microstructure morphology, densification, microhardness, grain refinement, pores, and coalescence (Pal et al., 2022).

4.1.2 Principles of Microwave Heating

The interaction of electromagnetic waves with materials is the foundation of microwave heating (Singh et al., 2018). With frequencies ranging from 300 MHz to 300 GHz and wavelengths ranging from 1 m to 1 mm, microwaves, a kind of electromagnetic radiation, can directly excite molecules inside materials (Chandrasekaran, Ramanathan, & Basak, 2012; Singh, Gupta, Jain, & Sharma, 2015). This excitation causes the material to heat quickly and uniformly by converting electromagnetic energy into thermal energy. The fundamental principles of microwave heating include the following:

4.1.2.1 Dielectric Heating

Dielectric materials may absorb microwave radiation and transform it into heat, including ceramics and polymers (Chandrasekaran et al., 2012). This is because the substance contains polar molecules or ions, which quickly reorient in response to the microwaves' oscillating electric field, causing friction and heat (Mishra & Sharma, 2016c).

4.1.2.2 Selective Heating

Materials with high dielectric loss coefficients can be heated selectively using microwave energy (Chandrasekaran et al., 2012; Mishra & Sharma, 2016c); consequently, materials with more significant concentrations of moisture, ions, or other polar molecules have the propensity to absorb microwave energy more readily and heat up more quickly than materials with lower dielectric loss factors.

4.1.2.3 Noncontact Heating

Unlike conventional heating techniques, microwave heating does not rely on conduction or convection to transfer heat. Instead, it immediately heats the materials with electromagnetic waves, allowing volumetric heating without contact (Bhattacharya & Basak, 2016).

4.1.2.4 Microwave–Material Interaction

The dielectric properties of materials, including the dielectric constant (permittivity) and the loss factor (dissipation factor), affect how microwaves interact. These characteristics determine a material's capacity to absorb microwave radiation and transform it into heat (Bhattacharya & Basak, 2016).

4.1.2.5 Dielectric Constant

The ability of a material to be polarized by an external electric field (such as that of microwaves) and store energy within it is called dielectric constant (ε') (Bhattacharya & Basak, 2016). Dielectric materials may absorb microwave radiation and transform it into heat.

4.1.2.6 Loss Factor

When a material is exposed to an electric field, the loss factor shows how much energy is lost as heat inside the substance. More excellent microwave absorption is seen in materials with more prominent loss factors, which results in more intense heating (Mishra & Sharma, 2016c).

4.1.2.7 Microwave Penetration Depth

Microwave penetration depth is the distance electromagnetic waves must travel through a substance before their energy is appreciably reduced (Bhattacharya & Basak, 2016; Mago, Bansal, Gupta, & Jain, 2020; Pal et al., 2022). Metal-based materials are heated by microwaves depending on the skin depth or the depth of penetration (δ).

The other definition of penetration depth or skin depth of the field is the distance from the material's surface at which the magnitude of field strength drops by a factor of $1/e$ (Mago et al., 2020; Mishra & Sharma, 2016b). Mathematically, it can be expressed as:

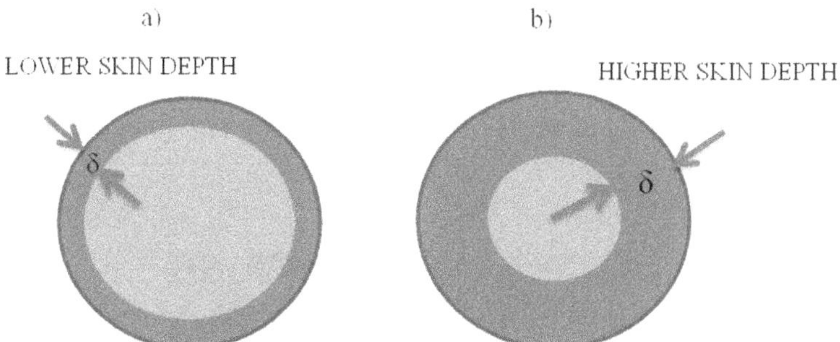

FIGURE 4.1 (a) Lower skin depth powder particles and (b) higher skin depth powder particles.

$$\delta = \frac{1}{\sqrt{\pi f \mu \sigma}} = 0.029\left(\rho_e \lambda_0\right)^{0.5} \tag{4.1}$$

(Mishra & Sharma, 2016b)

where f is the microwave frequency (2.45 GHz), μ is the magnetic permeability of the material (Hm^{-1}), σ is the electrical conductivity of the material ($S\ m^{-1}$), ρ_e is the electrical resistivity of the material (Ω m), and λ is the incident wavelength of microwave (m) (Mishra & Sharma, 2016b). The powder particles with higher skin depth melted earlier than those with lower skin depth particles. Figure 4.1a, b presents the lower and higher skin depth powder particles.

4.1.2.8 Microwave Hybrid Heating

A heating technique known as microwave hybrid heating (MHH) combines microwave radiation with other heating techniques, such as conduction, convection, or radiant heating (Tayier, Janasekaran, & Tai, 2022). To produce more effective and regulated heating of materials involves the simultaneous use of microwaves and another heat transfer method. In MHH, the material is selectively and voluminously heated with a lesser thermal gradient using microwaves.

Microwave processing of metal powder and metal-based materials depends upon the response of microwaves to the material. Two types of heating, direct and indirect, are used for bulk materials and metal powder casting (Mishra & Sharma, 2016c). In the direct heating method, materials made of metal are exposed to microwaves directly; in the indirect heating method, often referred to as microwave hybrid heating (MHH), susceptor materials are used to prevent direct microwave contact with metals. Both ways can process metal powder particles in a microwave. Due to the nonuniform temperature gradient throughout the material, the direct heating process results in poor microstructures on the surface and mechanical characteristics in metal powder compacts (Gupta & Wong, 2005). At the same time, microwave hybrid heating gives us a suitable microstructure with the refinement of grains. Many researchers in various fields successfully implemented the MHH technique for metallic powder materials processing (Gupta & Sharma, 2011; He, Wang, Wang, Liu, & Duan, 2022; Kaushal, Gupta, & Bhowmick, 2017b; Mishra & Sharma, 2016a; Pal et al., 2022; Singh et al., 2016, 2018). The complete mechanism used in this study in the form of graphic representations of several tools created for hybrid microwave heating is shown in Figure 4.2.

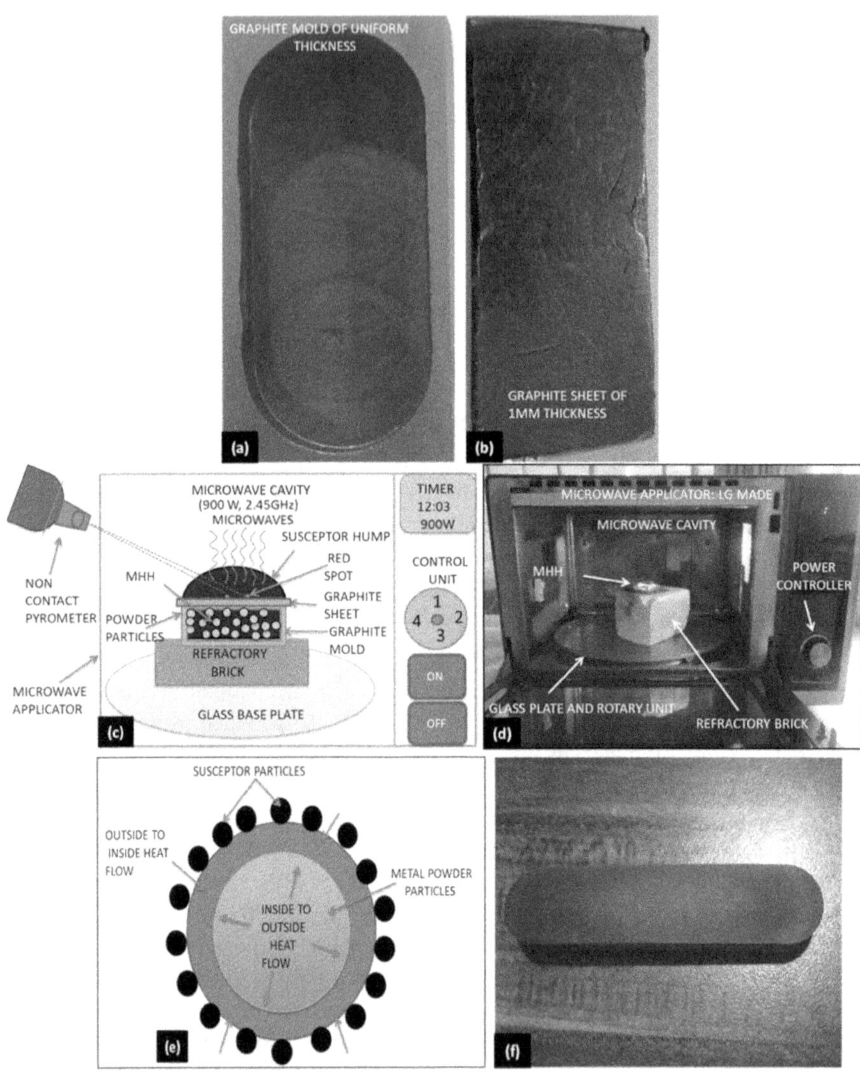

FIGURE 4.2 Tools for microwave process: (a) graphite mold cavity, (b) graphite sheet, (c) schematic view of microwave hybrid heating (MHH), (d) actual view of volumetric and selective MHH, (e) microwave hybrid heating (MHH) mechanism, (f) defect-free casting specimen.

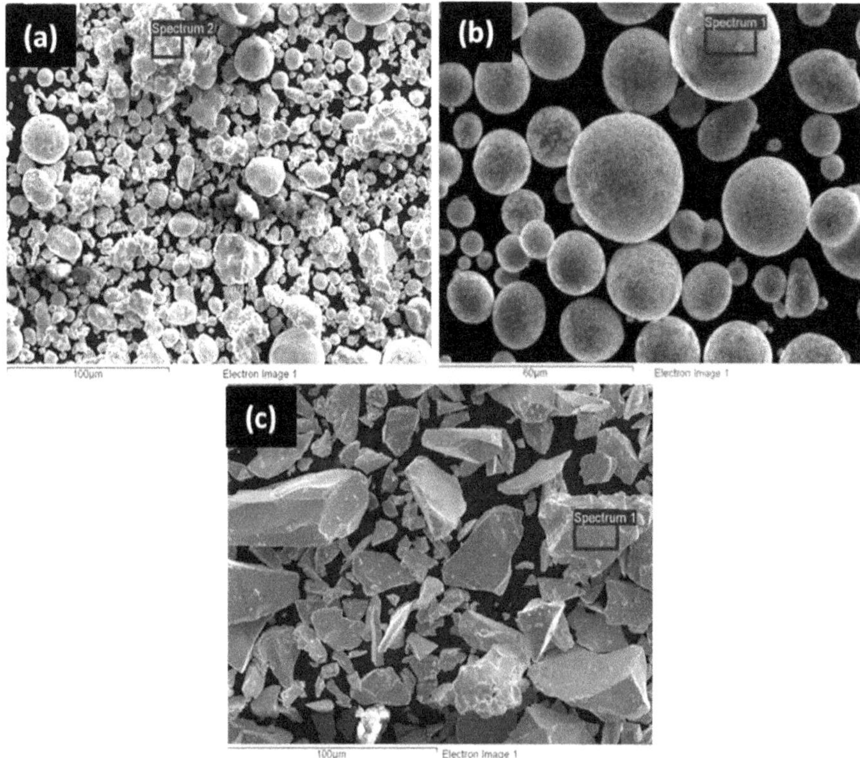

FIGURE 4.3 (a) SEM micrograph morphology of (a) ferrous-base alloy (SS316) powder, (b) Ni powder, and (c) Cr_3C_2.

Figure 4.2a, b presents the prepared graphite mold cavity for casting, having a uniform thickness of 2 mm and a separator sheet of 1 mm thickness used for covering the mold cavity to prevent contamination. Figure 4.2c, d shows the schematic and actual views of the complete MHH process with full descriptions. Figure 4.2e presents the selective and volumetric MHH, and Figure 4.2f shows the prepared casting specimen without any typical defects.

In this study, different powder materials have been used for preparing composite cast using microwave energy. The metallic powder materials morphology is shown in Figure 4.3, which indicates the spherical and irregular shapes of ferrous-based alloy (SS316) powder, the spherical morphology of Ni powders, and sharp-edged morphology of Cr_3C_2. The microstructure of the microwave-processed cast specimen is shown in Figure 4.4, which indicates the proper coalescence and refinement of grains of prepared samples.

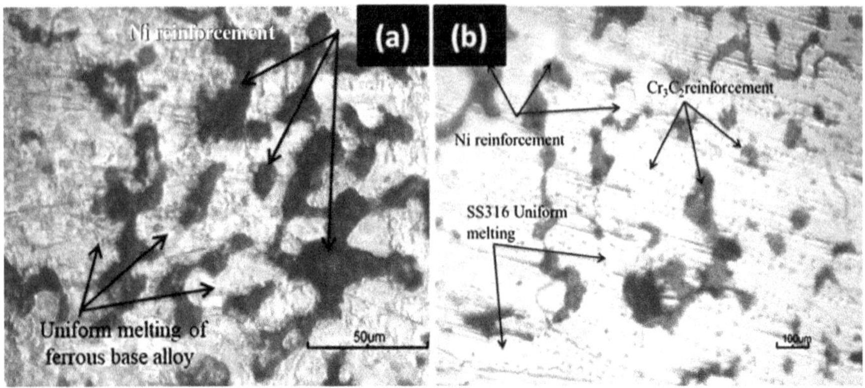

FIGURE 4.4 Optical micrograph morphology of prepared microwave cast samples in this study: (a) ferrous-base alloy (SS316)+10%Ni and (b) ferrous-base alloy+10%Ni +5%Cr_3C_2.

4.2 LITERATURE REVIEW

S.No	Findings
1	Bhoi, Singh, Pratap, and Jain (2019) claimed that microwave processing (MWP) of materials is one of its low-energy-consuming and sustainable manufacturing processes for the development of a large variety of materials in a solid, liquid, and semiliquid state.
2	Singh, Khanna, and Singh (2023) discussed the numerous applications of microwave energy in developing materials and processing technologies, its pros, and sustainability matters.
3	Mehta, Vasudev, and Jeyaprakash (2023) explained the basic properties of microwaves and their interaction mechanisms with the material's heating ability
4	Liu, Zhou, Xu, Han, and Zhou (2020) prepared SiC-reinforced aluminum metal matrix composites by microwave sintering. It was claimed that the relative density decreased with increasing content of reinforcement. The relative density was 98.43% and 96.14% at 5 vol% and 15 vol% of SiC/Al and the hardness of 130HV.
5	Kaushal, Gupta, and Bhowmick (2018a) the functionally graded clads of Ni-SiC material were developed on austenitic stainless steel (SS-304) substrate through a 2.45 GHZ domestic microwave applicator. The functionally graded clads were processed with power levels from 180 to 900 W. The exposure time of 900 W microwave power was varied from 300 s to 420 s. The maximum clad microhardness of 1020 ± 30 HV was achieved.
6	Singh et al. (2019) developed microwave-processed cast iron joints and improved the mechanical properties of the joint.
7	Singh, Gupta, and Jain (2019) successfully employed microwave heating for the melting and casting of EWAC (nickel-based) reinforced with silicon carbide particles (5–10%) and claimed 10% SiC-reinforced MMC revealed 2.4 times microhardness (980 ± 208$HV_{0.1}$), 42.58% (442 ± 20 MPa) higher tensile strength, and 2.54 times reduction in ductility than pure nickel-based casting
8	Singh et al. (2018) developed nickel-based MMC with 5–10% WC-8Co reinforcement using a microwave oven having a frequency of 2.45 GHz and power of 900 W and claimed the nickel powder melted within 25 minutes

S.No	Findings
9	Kaushal, Gupta, and Bhowmick (2018d) modified the surface characteristics of austenitic stainless steel using microwave-processed Ni/Cr_3C_2 composite cladding.
10	Mishra and Sharma (2018) prepared a microwave cast of the bulk copper using MHH at 2.45 GHz and power of 1400 W in a graphite mold, and the cast revealed 2–5% porosity and microhardness of 93 ± 20 HV, also the thin oxide layer plays a significant role for melting the bulk copper
11	Kaushal, Bohra, Gupta, and Jain (2020) cast Cu + 10% Mo, Cu + 30% Mo, and Cu + 50% Mo by volume percentage of powders using a microwave applicator cavity. They observed the existence of $Cu_6Mo_5O_{18}$, MoO_2, and $Cu_{64}O$ phases in XRD; the higher percentage of Mo inside Cu increases the electrical resistivity (5.68 ± 0.32) (10^{-8} Ωm) and hardness (120.8 ± 9 HV).
12	Ashwath and Xavior (2018) fabricated aluminum alloy (2900 and 2024) based metal matrix composites with alumina and silicon carbide as reinforcement having particle size 10 μm using a powder metallurgy route. Compressive strength and Rockwell hardness of the microwave-sintered MMC specimens (AA 2024 with 6 wt%) showed improved strength behavior, hardness, and stress-strain response.
13	Kaushal, Gupta, and Bhowmick (2017a) investigated a wear-resistant composite cladding of Ni-based +10% SiC on martensitic stainless steel (SS-420) through the MHH technique. The developed clad is uniformly developed and it is metallurgically bonded with the substrate. The Vicker's microhardness was 652 ± 90 HV. The clad surface showed good anti-wear properties.
14	Roy, Agrawal, Cheng, and Gedevanishvili (1999) successfully processed the sintering of metallic materials for the first time using microwave energy
15	Kaushal, Gupta, and Bhowmick (2021) It was revealed that microwave-processed functionally graded clad showed approximately 95% and 29% higher wear resistance than the SS-304 and Ni + 30%WC8Co-based single-layer clad
16	Pal et al. (2022) explained the complete mechanism of MHH and revealed that the microwave casting process could deliver clean and green energy at an optimum cost. The castings of iron-base alloy (SS316) powder and reinforced with 10 wt.% EWAC1004 (Ni-based) + 3 wt.% WC-12Co was successfully developed in the domestic microwave oven power of 900 W and a frequency of 2.45 GHz.
17	Kaushal, Gupta, and Bhowmick (2019) The hard phases contributed to the higher hardness of the FGC layer. The maximum value of microhardness occurred in the upper FGC layer, which was 1020±30 HV
18	Singh, Dinesh, Kaushal, and Singh (2021) investigated the effect of the power level on microwave processing materials manufacturing.
19	Kaushal, Singh, and Gupta (2022) explained a strategy of heating through microwaves to process the hybrid Ni-based composite containing 15% (WC-8Co) and 5% Mo clad on stainless steel SS 316L. Surface. The hybrid cladding was fruitfully prepared in 12–15 minutes of microwave exposure.
20	Kaushal, Singh, Gupta, and Jain (2019) developed metal matrix composite (MMC) claddings of Ni + 10% WC8Co + Cr_3C_2-based material on SS-316 L substrate using cost-effective microwave hybrid heating (MHH) used in severe wear conditions. The developed clads presented the refined microstructure with randomly dispersed reinforcement particles inside the Nickel matrix with the Co3W3C4, Cr7Ni3, NiC, NiW, W2C, Fe6W6C, Fe7C3, FeNi3 phases. The mean microhardness was observed as 550 ± 40 HV with better tribological properties than SS-316 L under different tribological conditions.

(*continued*)

S.No	Findings
21	Kaushal, Gupta, and Bhowmick (2018b) used a domestic microwave oven at 2.45 GHz and a power level of 180–900 W to prepare functionally graded clads (FGC) of Ni-Cr_3C_2 based composite powders with changing %age of Cr_3C_2 (0%–30% by wt.) on austenitic stainless steel (SS-304). The top FGC layer presents the maximum microhardness value of 576 ± 25 HV, which is 2.5 times more than the substrate.
22	Pal, Gupta, and Singh (2023b) studied the flexural behavior of composite castings under three-point loading conditions. The prepared sample's mechanical properties, such as flexural strength, peak load, strain hardness, and flexural modulus, are tested on Dak System UTM of load cell 10 kN as per ASTM 1161-18 standard.
23	Pal, Gupta, and Singh (2023a) explained microwave interaction with the metallic powder to optimize the processing time and temperature. A microstructure morphology of castings revealed the proper melting of matrix and EWAC (Ni-based) with an identical dispersion of Cr_7 C_3
24	Pal, Gupta, and Singh (2024) studied the dry sliding wear behavior of SS316-based composites cast in a domestic microwave oven with EWAC and WC-12Co reinforcements. The metallurgical, mechanical, and tribological properties of composites were improved than pure SS316 cast.

4.3 MICROWAVE-ASSISTED CASTING

The following procedure is used for the preparation of microwave-processed casting:

4.3.1 Preparation of Metallic Powders

The mechanically blended metallic composite powders of particle size 40 ± 15 μm with the required proportion are taken in a crucible of 100 g. The composite powders are preheated for 10–12 min at 200±10°C on convection mode in a domestic microwave. The warmed mixed powders are poured to the top into the graphite mold cavity. The mold is covered with a 1-mm-thick graphite sheet to protect it from environmental contamination. The mold is then placed on the refractory brick. The susceptor material (charcoal) is applied to the mold to surround it and form a good hump over it. The complete setup is placed in the domestic microwave on the glass plate of the rotary table.

4.3.2 Interactions of Microwaves with Metallic Powders

When the applicator is turned on, the susceptor powder immediately begins coupling with microwave (electromagnetic) radiations, converting these radiations into heat energy, as shown in Figure 4.2d. Due to the moisture in the susceptor (charcoal) powder, some sparks may also be observed from the charcoal for 1–5 seconds. The heat energy of electromagnetic radiation is created from the coupled microwave radiation, which is heated to the susceptor powder particles before being further transferred to the nearby powder particles. Initially, the metallic powder materials do not pair with microwaves because of the shallow skin depth of powder particles (Pal et al., 2022). The microwave starts to couple with the powder as soon as the temperature of the powder particles reaches a critical point due to heat being transferred from burning charcoal to the powder through conventional heating. When powder particles are heated (stimulated), their skin depth rises. As a result, there is a quick increase in the interaction of the metallic powder with electromagnetic radiation (microwaves), which raises the temperature and causes the composite powder to melt and cast (Pal et al., 2022). The principles of hybrid microwave heating (MHH) are used during microwave processing which had already been explained by earlier researchers in detail (Gupta & Sharma, 2011; Mishra & Sharma, 2016d; Pal et al., 2022; Sharma & Gupta, 2012).

Compared to conventional casting techniques, microwave-assisted casting has several benefits, such as heating the metal alloys quickly and effectively, reducing processing time, direct volumetric heating, and reducing heat losses, resulting in increased energy efficiency, improved material characteristics including smaller grain size, better mechanical performance, and improved microstructures. Resource usage and material waste are reduced thanks to careful process control and heating—the possibility of less harmful manufacturing processes that reduce emissions and environmental impact.

4.4 ADVANTAGES OF MICROWAVE-PROCESSED CASTING

Microwave-processed casting is a viable strategy for green manufacturing since it has several advantages over conventional casting techniques. The following are some of the main benefits of casting that have been microwave-processed.

4.4.1 Energy Efficiency

The processed substance receives an adequate and focused energy transfer by microwave heating. In contrast to conventional techniques, which rely on outside heat sources, volumetric heating during microwave processing directly warms the material (Bajpai, Singh, & Madaan, 2012; Gupta & Kumar, 2014; Kaushal, Gupta, & Bhowmick, 2018c; Xu et al., 2017). Because of the decreased energy use and shortened processing times, the casting process experiences significant energy savings (Lingappa & Amarendra, 2018). Compared to traditional castings, microwave processing can save up to 90% of the energy (Agrawal, 2006). High-temperature materials must be treated in a conventional furnace with a power rating of at least 2 kW. However, such materials were only processed in a microwave oven with a power rating of 900 W (Singh, Gupta, & Jain, 2021).

4.4.2 Reduced Processing Time

The microwaves' capacity for rapid and volumetric heating makes faster heat transfer within the material possible. Compared to traditional casting processes, this results in shorter processing cycles. Reduced processing times increase productivity while saving energy and accelerating manufacturing output (Panda, Singh, Upadhyaya, & Agrawal, 2006; Shashank et al., 2018; Yahaya, Izman, Idris, & Dambatta, 2015).

4.4.3 Enhanced Material Properties

Microwave heating results in homogeneous microstructures, improved densification (reduced porosity), uniform grain growth, higher modulus of rupture, high hardness, good flexural strength, high bending strength, reduced shrinkage distortion, and improved tribological properties (Agrawal, 2010; Agrawal, 2013; Kaushal, 2022; Singh et al., 2019; Kumar, Bhowmick, & Gupta, 2019; Pal et al., 2022; Singh et al., 2015).

4.4.4 Waste Reduction

Less material is wasted because microwave-treated casting enables accurate and controlled heating. Excess material and sprues can be reduced by focusing on the precise places that need to be heated. Additionally, the enhanced solidification process control allows for the fabrication of nearly net-shaped components minimizing the need for labor-intensive machining and material removal. Reducing waste not only lowers material costs but also helps make casting more environmentally and sustainably friendly (Agrawal, 2013; Kumar et al., 2019; Pal et al., 2022; Singh et al., 2015).

4.4.5 Environmental Benefits

Processing materials using a microwave benefits the environment by producing fewer emissions and pollutants. Microwave-processed casting can reduce greenhouse gas emissions compared to conventional casting techniques that rely on fossil fuel-based heating systems. The casting method leaves a minor carbon impact because of efficient energy use and quick processing times. Decreased material waste also helps with sustainable resource management and conservation (Pal et al., 2022). Regarding the energy economy, processing speed, improved material characteristics, waste reduction, and environmental benefits, microwave-treated casting offers several advantages. Due to these benefits, it is a viable and promising method for casting industry green manufacturing.

4.5 INDUSTRIAL APPLICATIONS AND CASE STUDIES

4.5.1 Microwave-Assisted Aluminum Casting

Microwave processing has produced encouraging outcomes in the aluminum casting sector. One case study involves the investment casting of aluminum alloys using microwave assistance. Compared to conventional approaches, the cycle times for shell drying and wax removal were dramatically shortened when using microwaves.

The procedure improved the cast aluminum components' mechanical qualities, dimensional correctness, and surface finish. This case study showed how aluminum casting operations can reduce energy use and increase productivity (Gajmal & Raut, 2019; Rajabi, Khodai, & Askari, 2014).

4.5.2 Microwave-Assisted Steel Casting

To increase the effectiveness and quality of steel casting, which is a crucial process in many sectors, microwave-assisted approaches have been investigated. Steel continuous casting with microwave assistance was the subject of the case study. The procedure increased microstructural uniformity in the cast steel products, eliminated flaws, and accelerated solidification by localized heating of the mold and the molten steel with microwaves. Utilizing microwave processing in steel casting has benefits in terms of shorter cycle times, improved product quality, and energy savings (Pal et al., 2022; Ram et al., 2022).

4.5.3 Microwave-Assisted Titanium Casting

Due to its high melting point and reactivity, casting with titanium presents particular difficulties. Applications for titanium casting have been looked into using microwave processing. Microwave-assisted investment casting for titanium alloys was proven in a case study. Microwaves were used to remove wax and dry the shell, which shortened cycle times and enhanced surface quality. Additionally, the method allowed for perfect temperature control, reducing the danger of titanium oxidation and producing cast titanium components of the highest caliber (Akbarpour, Alipour, Najafi, Ebadzadeh, & Kim, 2021; Choy et al., 2015).

4.5.4 Microwave-Assisted Non-Ferrous Metal Casting

Additionally, casting different non-ferrous metals, such as copper, brass, and bronze, has found use for microwave processing. A microwave-assisted sand casting of copper alloys was the subject of a case study that showed how the technique shortened mold drying periods and increased the dimensional accuracy of the cast copper components. Microwaves' effective and uniform heating improved microstructural characteristics and fewer flaws in the finished cast goods. In these sectors, microwave-assisted non-ferrous metal casting has the potential to reduce waste, save energy, and enhance product quality (Mishra & Sharma, 2018; Mizuno, Kosai, & Yamasue, 2021).

4.6 CHALLENGES AND FUTURE DIRECTIONS

4.6.1 Material Selection and Compatibility

Choosing appropriate materials is one of the difficulties in casting that has undergone microwave processing. Materials' interactions with microwaves are influenced by their dielectric characteristics, and not all materials can be heated effectively. The behavior of various materials during microwave processing conditions must be better understood to find suitable alloys or composites for a given casting application (Samyal, Bagha, & Bedi, 2020).

4.6.2 Equipment Design and Optimization

To get the best results, microwave heating systems and casting machinery (as shown in Figure 4.2a–f) must be designed carefully. Creating effective microwave applicators, sensors, and feedback control systems is crucial for accurate and consistent heating. Process control factors must be continuously monitored and optimized to guarantee consistent and dependable casting results, including power density, heating rate, and temperature profiles (Mishra & Sharma, 2016a; Pietrenko-Dabrowska & Koziel, 2019).

4.6.3 Standardization

Process, tool, and quality control measure standardization are required to encourage the general use of microwave-treated casting. Setting up industry standards and regulations will guarantee uniform and repeatable outcomes across various manufacturing sites. To provide standardized standards for microwave-processed casting processes, cooperation between researchers, business leaders, and governing organizations is essential (Samyal et al., 2020).

4.6.4 Integration with Other Green Manufacturing Technologies

The sustainability of casting operations can be further improved by combining microwave processing with other environmentally friendly manufacturing techniques. For instance, using recycled materials or combining microwave heating with additive manufacturing processes can help reduce waste and conserve resources. An exciting area for further study and development is finding ways microwave processing and other environmentally friendly manufacturing techniques might work together. Although casting that has been microwave-processed may have some positive environmental effects, it is crucial to perform thorough life-cycle analyses to see how it will affect the ecosystem as a whole. Assessments should consider variables, including energy use, emissions, waste production, and raw material usage. A more precise evaluation of the sustainability of microwave-treated casting can be obtained by evaluating its environmental performance and contrasting it with traditional techniques (Mondal, Upadhyaya, & Agrawal, 2008).

4.6.5 Scalability and Cost-Effectiveness

Scalability and cost-effectiveness issues become more critical as microwave-processed casting methods develop. Creating scalable, economical systems that can be incorporated into current production infrastructures is crucial. Microwave-processed casting will be more scalable and economical because of improvements in microwave technology and optimization of process parameters and equipment design (Pal et al., 2022).

4.6.6 Industry Adoption and Knowledge Transfer

It takes information transfer and awareness-raising initiatives to help the industry adopt microwave-processed casting. Collaborations between sectors, workshops, and training initiatives can promote microwave processing's advantages, difficulties, and best practices. The adoption and application of microwave-processed casting techniques will be accelerated by encouraging information sharing and collaboration between academic institutions, research organizations, and industrial stakeholders.

The ongoing development and implementation of microwave-treated casting as a sustainable and effective manufacturing technology will be made possible by addressing these issues and considering potential possibilities. Microwave processing has the potential to transform the casting industry, resulting in greener, more efficient, and higher-quality casting processes with continued research and developments.

4.6.7 Environmental Impact Assessment

To understand the overall sustainability of microwave-processed casting and to pinpoint areas for development, it is essential to evaluate its environmental impact. As you prepare to undertake an environmental impact assessment, keep the following points in mind:

a. **Energy Consumption:** Consider the power needs for microwave generators, applicators, and ancillary equipment when assessing the energy consumption of microwave processing (Singh et al., 2021). To evaluate possible energy savings, compare it to the energy usage of traditional casting techniques.
b. **Emissions:** Examine the emissions of greenhouse gases, particulate matter, and volatile organic compounds produced during the manufacture and usage of microwave processing equipment. To establish whether these emissions have any positive or negative effects on the environment, compare them to those produced by traditional casting techniques.

Analyze the utilization of raw resources, such as metal alloys, mold materials, and consumables, in microwave-treated casting. Take into account the environmental impact of their manufacturing, transportation, and disposal. Examine the possibility of employing recycled or naturally derived materials.

c. **Waste Generating:** Consider the waste materials produced during the mold-making and wax-removing stages of the casting cycle, as well as any surplus or flawed castings. Analyze waste management techniques, such as recycling, reuse, or proper disposal, to reduce environmental impact.
d. **Water Consumption:** Consider how much water is used during the microwave processing of casting, such as for cooling systems or the creation of

mold. Find chances to reduce water use through process improvement or different cooling techniques.

e. **Life-Cycle Assessment:** From the extraction of raw materials to the end-of-life stage, conduct a life-cycle assessment to examine the overall environmental impact of microwave-treated casting. Consider each stage's effects on the environment, including energy use, emissions, trash production, and resource depletion, and analyze the comparative effects of microwave-processed and traditional casting techniques on the environment.
f. **Continuous Improvement:** Find areas for process improvement and environmental performance enhancement using the results of the environmental impact assessment. To further lessen the environmental impact of microwave-treated casting, consider alternate materials, equipment upgrades, or process modifications.

Manufacturers may learn more about the sustainability of microwave-processed casting and put initiatives in place to lessen its environmental impact by completing a thorough environmental impact assessment. This analysis offers a foundation for making decisions, streamlining processes, and creating sustainable manufacturing procedures.

4.7 CONCLUSION

As a green production method, microwave-processed casting has much potential in the casting sector. The basics, benefits, casting techniques, case studies, difficulties, and potential future directions of microwave-processed casting have been discussed in this chapter. The following are the main conclusions from the discussion:

a. Compared to conventional casting techniques, microwave-treated casting has many advantages, including shortened processing time, improved material characteristics, less waste, and favorable environmental effects. These benefits help to boost sustainability, reduce costs, and increase production.
b. Microwave processing is advantageous for several conventional and other casting processes, including investment casting, continuous casting, die casting, sand casting, and powder metallurgy. These approaches use microwaves to increase material quality, process efficiency, and overall product quality.
c. The casting of non-ferrous metals, steel, titanium, and aluminum has shown the effectiveness of microwave processing. According to these studies, casting techniques now have shorter cycle durations, better dimensional accuracy, superior mechanical characteristics, and use more energy efficiently.
d. Microwave-treated casting still has problems despite its many benefits. These consist of the choice of materials, the design of the machinery and process management, standardization, integration with other green manufacturing technologies, the evaluation of the environmental impact, scalability, and industry adoption. Continuous research, technical developments, cooperation, and knowledge sharing are necessary to address these issues.

e. An environmental impact assessment must be conducted to determine if casting that has undergone microwave processing is sustainable. In addition to conducting comparison studies with traditional casting techniques, checks should consider energy usage, emissions, raw materials, waste generation, and water consumption. The results of these analyses offer guidance for process improvement, waste minimization, and environmental performance enhancement.

In the end, microwave-processed casting can potentially revolutionize the casting sector by providing practical, environmentally friendly, and high-quality manufacturing processes. Microwave-processed casting has the potential to become a widely used green manufacturing method with additional study, technological development, and industry cooperation, helping to create a casting industry that is more environmentally friendly and sustainable.

REFERENCES

Agrawal, D. (2006). Microwave sintering, brazing and melting of metallic materials. In *Sohn International Symposium: Advanced Processing of Metals and Materials – Proceedings of the International Symposium* (pp. 183–192). (2006 TMS Fall Extraction and Processing Division: Sohn International Symposium; Vol. 4).

Agrawal, D. (2010). Latest global developments in microwave materials processing. *Materials Research Innovations, 14*(1), 3–8. https://doi.org/10.1179/143307510X12599329342926

Agrawal, D. (2013). Microwave sintering of metal powders. *Advances in Powder Metallurgy: Properties, Processing and Applications*, 361–379. https://doi.org/10.1533/9780857098900.3.361

Akbarpour, M. R., Alipour, S., Najafi, M., Ebadzadeh, T., & Kim, H. S. (2021). Microstructural characterization and enhanced hardness of nanostructured Ni3Ti–NiTi (B2) intermetallic alloy produced by mechanical alloying and fast microwave-assisted sintering process. *Intermetallics, 131*, 107119. https://doi.org/10.1016/j.intermet.2021.107119

Al-Shetwi, A. Q. (2022). Sustainable development of renewable energy integrated power sector: Trends, environmental impacts, and recent challenges. *Science of the Total Environment, 822*, 153645. https://doi.org/10.1016/j.scitotenv.2022.153645

Ashwath, P., & Xavior, M. A. (2018). Effect of ceramic reinforcements on microwave sintered metal matrix composites. *Materials and Manufacturing Processes, 33*(1), 7–12. https://doi.org/10.1080/10426914.2016.1244851

Bajpai, P. K., Singh, I., & Madaan, J. (2012). Joining of natural fiber reinforced composites using microwave energy: Experimental and finite element study. *Materials and Design, 35*, 596–602. https://doi.org/10.1016/j.matdes.2011.10.007

Bhattacharya, M., & Basak, T. (2016). A review on the susceptor assisted microwave processing of materials. *Energy, 97*, 306–338. https://doi.org/10.1016/j.energy.2015.11.034

Bhoi, N. K., Singh, H., Pratap, S., & Jain, P. K. (2019). Microwave material processing: a clean, green, and sustainable approach. In *Sustainable Engineering Products and Manufacturing Technologies* (pp. 3–23). https://doi.org/10.1016/B978-0-12-816564-5.00001-3

Buchbinder, D., Schleifenbaum, H., Heidrich, S., Meiners, W., & Bültmann, J. (2011). High power selective laser melting (HP SLM) of aluminum parts. *Physics Procedia, 12*, 271–278. https://doi.org/10.1016/j.phpro.2011.03.035

Chandrasekaran, S., Ramanathan, S., & Basak, T. (2012). Microwave material processing-a review. *AIChE Journal, 58*(2), 330–363. https://doi.org/10.1002/aic.12766

Choy, M.-T., Tang, C.-Y., Chen, L., Law, W.-C., Tsui, C.-P., & Lu, W. W. (2015). Microwave assisted-in situ synthesis of porous titanium/calcium phosphate composites and their in vitro apatite-forming capability. *Composites Part B: Engineering*, *83*, 50–57. https://doi.org/10.1016/j.compositesb.2015.08.046

Gajmal, S., & Raut, D. N. (2019). A review of opportunities and challenges in microwave assisted casting. *Recent Trends in Production Engineering*, *2*(1), 1–17.

Gupta, D., & Kumar, A. (2014). Microwave cladding: A new approach in surface engineering. *Journal of Manufacturing Processes*, *16*(2), 176–182. https://doi.org/10.1016/j.jmapro.2014.01.001

Gupta, D., & Sharma, A. K. (2011). Development and microstructural characterization of microwave cladding on austenitic stainless steel. *Surface and Coatings Technology*, *205*(21–22), 5147–5155. https://doi.org/10.1016/j.surfcoat.2011.05.018

Gupta, M., & Wong, W. L. E. (2005). Enhancing overall mechanical performance of metallic materials using two-directional microwave assisted rapid sintering. *Scripta Materialia*, *52*(6), 479–483. https://doi.org/10.1016/j.scriptamat.2004.11.006

Hashim, J., Looney, L., & Hashmi, M. S. (1999). Metal matrix composites: production by the stir casting method. *Journal of Materials Processing Technology*, *92–93*, 1–7. https://doi.org/10.1016/S0924-0136(99)00118-1

He, Z., Wang, Z., Wang, D., Liu, X., & Duan, B. (2022). Microstructure and mechanical properties of porous NiTi alloy prepared by integration of gel-casting and microwave sintering. *Materials*, *15*(20), 7331. https://doi.org/10.3390/ma15207331

Jones, S., & Yuan, C. (2003). Advances in shell moulding for investment casting. *Journal of Materials Processing Technology*, *135*(2–3), 258–265. https://doi.org/10.1016/S0924-0136(02)00907-X

Kaushal, S. (2022). Microstructure and tribological characterization of composite castings developed through in-situ microwave hybrid heating. *International Journal of Metalcasting*, *16*(4), 2150–2161. https://doi.org/10.1007/s40962-022-00757-1

Kaushal, S., Bohra, S., Gupta, D., & Jain, V. (2020). On processing and characterization of Cu–Mo-based castings through microwave heating. *International Journal of Metalcasting*. https://doi.org/10.1007/s40962-020-00481-8

Kaushal, S., Gupta, D., & Bhowmick, H. (2017a). Investigation of dry sliding wear behavior of Ni–SiC microwave cladding. *Journal of Tribology*, *139*(4). https://doi.org/10.1115/1.4035147

Kaushal, S., Gupta, D., & Bhowmick, H. (2017b). On microstructure and wear behavior of microwave processed composite clad. *Journal of Tribology*, *139*(6). https://doi.org/10.1115/1.4035844

Kaushal, S., Gupta, D., & Bhowmick, H. (2018a). On development and wear behavior of microwave-processed functionally graded Ni-SiC clads on SS-304 substrate. *Journal of Materials Engineering and Performance*, *27*(2), 777–786. https://doi.org/10.1007/s11665-017-3110-z

Kaushal, S., Gupta, D., & Bhowmick, H. (2018b). On processing of Ni-Cr_3C_2 based functionally graded clads through microwave heating. *Materials Research Express*, *5*(6), 066405. https://doi.org/10.1088/2053-1591/aac805

Kaushal, S., Gupta, D., & Bhowmick, H. (2018c). On processing of Ni-WC based functionally graded composite clads through microwave heating. *Materials and Manufacturing Processes*, *33*(8), 822–828. https://doi.org/10.1080/10426914.2017.1401724

Kaushal, S., Gupta, D., & Bhowmick, H. (2018d). On surface modification of austenitic stainless steel using microwave processed Ni-Cr_3C_2 composite cladding. *Surface Engineering*, *34*(11), 809–817. https://doi.org/10.1080/02670844.2017.1362808

Kaushal, S., Gupta, D., & Bhowmick, H. (2019). On processing and flexural behaviour of functionally graded clads developed through microwave irradiation. *Materials Research*

Express, *6*(7), 076405. https://doi.org/10.1088/2053-1591/ab11ea(Not accessible as of [2024/06/14])

Kaushal, S., Singh, D., Gupta, D., & Jain, V. (2019). Wear resistance improvement of austenitic 316 L steel by microwave-processed composite clads. *Journal of Tribology*, *141*(4). https://doi.org/10.1115/1.4042273

Kaushal, S., Gupta, D., & Bhowmick, H. (2021). Wear behavior of microwave-processed Ni-WC8Co-based functionally graded materials. *Proceedings of the Institution of Mechanical Engineers, Part L: Journal of Materials: Design and Applications*, *235*(5), 1036–1045. https://doi.org/10.1177/1464420720988119

Kaushal, S., Singh, S., & Gupta, D. (2022). Processing strategy for high strength Ni-based hybrid composite clad on SS 316L steel through microwave heating. *Proceedings of the Institution of Mechanical Engineers, Part B: Journal of Engineering Manufacture*, *236*(3), 190–203. https://doi.org/10.1177/09544054211021360

Kumar, R., Bhowmick, H., & Gupta, D. (2019). Progress on the development of aluminium metal matrix composites as an alternative piston material- A review. *Journal of Emerging Technologies and Innovative Research*, *6*(5), 414–419.

Kushwaha, G. S., & Sharma, N. K. (2016). Green initiatives: a step towards sustainable development and firm's performance in the automobile industry. *Journal of Cleaner Production*, *121*, 116–129. https://doi.org/10.1016/j.jclepro.2015.07.072

Lingappa, S. M., & Amarendra, M. S. S. H. J. (2018). Melting of bulk non-ferrous metallic materials by microwave hybrid heating (MHH) and conventional heating: a comparative study on energy consumption. *Journal of the Brazilian Society of Mechanical Sciences and Engineering*, *40*(1), 1–11. https://doi.org/10.1007/s40430-017-0921-7

Lingappa, S. M., Srinath, M. S., & Amarendra, H. J. (2018). Melting of bulk non-ferrous metallic materials by microwave hybrid heating (MHH) and conventional heating: a comparative study on energy consumption. *Journal of the Brazilian Society of Mechanical Sciences and Engineering*, *40*(1), 1. https://doi.org/10.1007/s40430-017-0921-7

Liu, J., Zhou, B., Xu, L., Han, Z., & Zhou, J. (2020). Fabrication of SiC reinforced aluminium metal matrix composites through microwave sintering. *Materials Research Express*, *7*(12), 125101. https://doi.org/10.1088/2053-1591/abc8bf

Mago, J., Bansal, S., Gupta, D., & Jain, V. (2020). Investigation of microwave processing parameters on development of Ni-Cr_3C_2 composite clad and their characterization. *Metallurgical and Materials Transactions A*, *51*(8), 4288–4300. https://doi.org/10.1007/s11661-020-05832-y

Mehta, A., Vasudev, H., & Jeyaprakash, N. (2023). Role of sustainable manufacturing approach: microwave processing of materials. *International Journal on Interactive Design and Manufacturing (IJIDeM)*. https://doi.org/10.1007/s12008-023-01318-4

Mishra, R. R., & Sharma, A. K. (2016a). A review of research trends in microwave processing of metal-based materials and opportunities in microwave metal casting. *Critical Reviews in Solid State and Materials Sciences*, *41*(3), 217–255. https://doi.org/10.1080/10408436.2016.1142421

Mishra, R. R., & Sharma, A. K. (2016b). A review of research trends in microwave processing of metal-based materials and opportunities in microwave metal casting. *Critical Reviews in Solid State and Materials Sciences*, *41*(3), 217–255. https://doi.org/10.1080/10408436.2016.1142421. https://doi.org/10.1080/10408436.2016.1142421

Mishra, R. R., & Sharma, A. K. (2016c). Microwave–material interaction phenomena: Heating mechanisms, challenges and opportunities in material processing. *Composites Part A: Applied Science and Manufacturing*, *81*, 78–97. https://doi.org/10.1016/j.compositesa.2015.10.035

Mishra, R. R., & Sharma, A. K. (2016d). On mechanism of in-situ microwave casting of aluminium alloy 7039 and cast microstructure. *Materials & Design*, *112*, 97–106. https://doi.org/10.1016/j.matdes.2016.09.041

Mishra, R. R., & Sharma, A. K. (2018). Experimental investigation on in-situ microwave casting of copper. *IOP Conference Series: Materials Science and Engineering*, *346*, 012052. https://doi.org/10.1088/1757-899X/346/1/012052

Mizuno, N., Kosai, S., & Yamasue, E. (2021). Microwave-based extractive metallurgy to obtain pure metals: A review. *Cleaner Engineering and Technology*, *5*, 100306. https://doi.org/10.1016/j.clet.2021.100306

Mondal, A., Upadhyaya, A., & Agrawal, D. (2008). Microwave and conventional microwave and conventional sintering of premixed and prealloyed tungsten heavy alloys. *Materials Science and Technology (MS&T)*, (January), 2502–2515.

Pal, J., Gupta, D., & Singh, T. P. (2022). Processing and characterization of SS316 based metal matrix composite casting through microwave hybrid heating. *Proceedings of the Institution of Mechanical Engineers, Part C: Journal of Mechanical Engineering Science*, *236*(20), 10508–10527. https://doi.org/10.1177/09544062221104443

Pal, J., Gupta, D., & Singh, T. P. (2023a). Development of iron-based composite materials through electromagnetic radiations in a domestic microwave oven. *Proceedings of the Institution of Mechanical Engineers, Part C: Journal of Mechanical Engineering Science*. https://doi.org/10.1177/09544062231181832

Pal, J., Gupta, D., & Singh, T. P. (2023b). Influence of tungsten carbide-cobalt reinforcement on flexural behavior of iron-based composite developed through microwave. *Proceedings of the Institution of Mechanical Engineers, Part C: Journal of Mechanical Engineering Science*. https://doi.org/10.1177/09544062231181826

Pal, J., Gupta, D., & Singh, T. P. (2024). Influence of WC-12Co reinforcement on sliding wear performance of stainless steel-based MMC processed in a microwave oven. *Proceedings of the Institution of Mechanical Engineers, Part C: Journal of Mechanical Engineering Science*, 09544062241228389. https://doi.org/10.1177/09544062241228389

Panda, S. S., Singh, V., Upadhyaya, A., & Agrawal, D. (2006). Sintering response of austenitic (316L) and ferritic (434L) stainless steel consolidated in conventional and microwave furnaces. *Scripta Materialia*, *54*(12), 2179–2183. https://doi.org/10.1016/j.scriptamat.2006.02.034

Pietrenko-Dabrowska, A., & Koziel, S. (2019). Numerically efficient algorithm for compact microwave device optimization with flexible sensitivity updating scheme. *International Journal of RF and Microwave Computer-Aided Engineering*, *29*(7), e21714. https://doi.org/10.1002/mmce.21714

Rajabi, M., Khodai, M. M., & Askari, N. (2014). Microwave-assisted sintering of Al–ZrO2 nano-composites. *Journal of Materials Science: Materials in Electronics*, *25*(10), 4577–4584. https://doi.org/10.1007/s10854-014-2206-6

Ram, V. K., Nandwani, S., Vardhan, S., Bahl, S., Samyal, R., & Bagha, A. K. (2022). Microwave casting of stainless still through microwave hybrid heating. *IOP Conference Series: Materials Science and Engineering*, *1248*(1), 012047. https://doi.org/10.1088/1757-899X/1248/1/012047

Roy, R., Agrawal, D., Cheng, J., & Gedevanishvili, S. (1999). Full sintering of powdered-metal bodies in a microwave field. *Nature*, *399*(6737), 668–670. https://doi.org/10.1038/21390

Sama, S. R., Wang, J., & Manogharan, G. (2018). Non-conventional mold design for metal casting using 3D sand-printing. *Journal of Manufacturing Processes*, *34*, 765–775. https://doi.org/10.1016/j.jmapro.2018.03.049

Samyal, R., Bagha, A. K., & Bedi, R. (2020). The casting of materials using microwave energy: A review. *Materials Today: Proceedings*, *26*, 1279–1283. https://doi.org/10.1016/j.matpr.2020.02.255

Sharma, A. K., & Gupta, D. (2012). On microstructure and flexural strength of metal–ceramic composite cladding developed through microwave heating. *Applied Surface Science, 258*(15), 5583–5592. https://doi.org/10.1016/j.apsusc.2012.02.019

Shashank, M. L. (2018). Casting of commercially available bulk copper by microwave melting process and its characterization *IOP Conference Series: Material Science and Engineering, 330*, 012087. DOI:10.1088/1757-899X/330/1/012087

Silvestroni, L., & Sciti, D. (2010). Sintering behavior, microstructure, and mechanical properties: a comparison among pressureless sintered ultra-refractory carbides. *Advances in Materials Science and Engineering*, 1–11. https://doi.org/10.1155/2010/835018

Singh, C., Khanna, V., & Singh, S. (2023). Sustainability of microwave heating in materials processing technologies. *Materials Today: Proceedings, 73*, 241–248. https://doi.org/10.1016/j.matpr.2022.07.216

Singh, G., Dinesh, Kaushal, S., & Singh, S. (2021). *Effect of Power Level on the Processing of Ni-Based Casting Through Microwave Heating*. https://doi.org/10.1007/978-981-15-5519-0_33

Singh, R., Singh, S., & Hashmi, M. S. J. (2016). Investment casting. In *Reference Module in Materials Science and Materials Engineering*. https://doi.org/10.1016/B978-0-12-803581-8.04163-1

Singh, S., Gupta, D., & Jain, V. (2016). Novel microwave composite casting process: Theory, feasibility and characterization. *JMADE, 111*, 51–59. https://doi.org/10.1016/j.matdes.2016.08.071

Singh, S., Gupta, D., & Jain, V. (2018). Processing of Ni-WC-8Co MMC casting through microwave melting. *Materials and Manufacturing Processes, 33*(1), 26–34. https://doi.org/10.1080/10426914.2017.1291954

Singh, S., Gupta, D., & Jain, V. (2019). Microwave processing and characterization of nickel powder based metal matrix composite castings. *Materials Research Express, 6*(8), 0865b1. https://doi.org/10.1088/2053-1591/ab2138

Singh, S., Gupta, D., & Jain, V. (2021). Fabricating in situ powdered nickel–alumina metal matrix composites through microwave heating process: a sustainable approach. *International Journal of Metalcasting, 15*(3), 969–982. https://doi.org/10.1007/s40962-020-00536-w

Singh, S., Gupta, D., Jain, V., & Sharma, A. K. (2015). Microwave processing of materials and applications in manufacturing industries: a review. *Materials and Manufacturing Processes, 30*(1), 1–29. https://doi.org/10.1080/10426914.2014.952028

Singh, S., Singh, P., Gupta, D., Jain, V., Kumar, R., & Kaushal, S. (2019). Development and characterization of microwave processed cast iron joint. *Engineering Science and Technology, an International Journal, 22*(2), 569–577. https://doi.org/10.1016/j.jestch.2018.10.012

Tayier, W., Janasekaran, S., & Tai, V. C. (2022). Microwave hybrid heating (MHH) of Ni-based alloy powder on Ni and steel-based metals – A review on fundamentals and parameters. *International Journal of Lightweight Materials and Manufacture, 5*(1), 58–73. https://doi.org/10.1016/j.ijlmm.2021.10.002

Xu, L., Peng, J., Bai, H., Srinivasakannan, C., Zhang, L., & Wu, Q. (2017). Application of microwave melting for the recovery of tin powder. *Engineering, 3*(3), 423–427. https://doi.org/10.1016/J.ENG.2017.03.006

Yahaya, B., Izman, S., Idris, M. H., & Dambatta, M. S. (2015). Effects of activated charcoal on dewaxing time in microwave hybrid heating. *Procedia CIRP, 26*, 467–472. https://doi.org/10.1016/j.procir.2014.07.039

5 An Introduction of Green Energy-Based Welding Process

Sanjeev Kumar

Abbreviation	State
FSW	Friction Stir Welding
FSWed	Friction Stir Welded
TRS	Tool Rotational Speed
TTS	Tool Traverse Speed
TTA	Tool Tilt Angle
Al-Li	Aluminium Lithium
NZ	Nugget Zone
TMAZ	Thermo Mechanical Heat Affected Zone

5.1 INTRODUCTION

The joining techniques provide a more ecologically responsible option to conventional welding procedures that rely on fossil fuels; green energy-based welding processes are growing in popularity. Following are some instances of welding techniques based on green energy:

a. **Solar Welding**
   This type of welding uses the sun's energy to produce a powerful arc that melts the metal being welded. The welding of pipes, tanks, and other metal structures is one of the many uses for solar welding.
b. **Using Renewable Energy for Electric Arc Welding**
   A common welding technique is electric arc welding, which commonly uses grid energy or diesel generators for joining. However, other energy sources like solar or wind power can also be used to power electric arc welding.
c. **Friction Stir Welding**
   Friction stir welding (FSW) is a solid-state joining process that joins the metals, nonmetals, polymer, steel, metal matrix composite, and so on without melting them. Instead, the metal is heated and made softer through friction, which is then pressed together to form a connection. Lightweight materials

DOI: 10.1201/9781003449225-5

such as aluminum, magnesium, Al-Li alloy, and so on are joined using this technique (Kumar et al. 2021).

d. **Using Renewable Energy for Laser Welding**
A laser beam is used to melt and combine metal components during the high-precision welding technique known as laser welding. Laser welding can be fueled by renewable energy sources like solar or wind power, just like electric arc welding.

e. **Microwave Joining**
The microwave joining technique allows materials to be bonded together using electromagnetic waves within the range of microwave frequency in a dedicated microwave chamber. When the microwaves are turned on, they enter the materials and interact with the molecules therein. Rapid molecular vibrations caused by this contact cause heat to be produced by friction (Kaushal et al. 2017, 2018a). Materials like certain polymers and plastics (polypropylene, polyethylene, and polycarbonate have higher dielectric loss tangent values), composite materials (fiber-reinforced composites and composite laminates contain dielectric components responsive to microwave energy), ceramic materials (include oxides such as Al_2O_3 and $Al_6Si_2O_{13}$, Si_3N_4, and SiC) (Silberglitt et al. 1993), certain types of wood (higher moisture content, which can be effectively joined using microwave energy due to its dielectric properties), and certain glasses (especially those with high lead content) can be joined using the microwave joining process (Kaushal et al. 2019a, 2020).

The use of green energy-based welding procedures is expected to increase as the world transitions to a low-carbon economy since they provide a safer and more environmentally friendly alternative to conventional welding methods. It is frequently referred to as a "green" welding method based on different aspects as given below (Kumar et al. 2022c, 2023):

a. **Energy Efficiency**
Compared to typical welding methods, FSW uses less energy because the metal being welded together is not melted during the welding process.

b. **Reduced Emissions**
When compared to conventional welding methods, FSW emits less pollution because it doesn't employ filler metals or gases. Additionally, FSW does not emit any fumes or smoke, which decreases environmental pollution and makes it the more secure welding technique for the operator.

c. **Reduced Waste**
FSW uses a solid-state joining process to form a continuous, defect-free junction, which results in less waste than other welding procedures. As a result, there are less scrap parts and the manufacturing process is more effective.

d. **Recyclability**
FSW produces joints that are extremely durable and don't need any further coatings or treatments. As a result, there will be less waste and environmental impact when the linked pieces are easily recyclable at the end of their life cycle.

5.2 MICROWAVE JOINING PROCESS

Microwave joining is a state-of-the-art technology that facilitates the bonding of materials using electromagnetic waves in the microwave frequency range. Microwave joining uses the intensity of electromagnetic waves in the microwave frequency range compared to more traditional techniques like welding and soldering. These waves generate localized heating to occur within the materials, which causes a solid bond to develop. The materials to be bonded are first placed within a specialized microwave chamber (Kaushal et al. 2019b, 2021, 2022). When the microwaves are turned on, they enter the materials and interact with the molecules therein. Rapid molecular vibrations caused by this contact cause heat to be produced by friction. The materials can fuse together as the temperature rises because they will eventually reach their melting or softening thresholds. When the process is finished, the materials cool down, and a solid and lasting bond is created between them.

5.2.1 Applications of Microwave Joining Process

A sophisticated method used to forge solid connections between various materials is microwave joining. Instead of using traditional techniques like soldering and welding, microwave joining harnesses. Due to its capacity to forge effective and powerful bindings between various materials, microwave joining has found a wide range of useful applications across numerous industries. Key applications include the following:

a. **Aerospace and Defense Industry**
Composites and other lightweight materials are widely used in the aerospace and defense sectors. In crucial components like airplane panels and engine parts, microwave joining is used to form trusty bonds. Because it may lessen deformation during bonding, aircraft components' structural integrity is guaranteed.
b. **Automotive Industry**
The ability of microwave joining to join lightweight materials like aluminum and carbon fiber composites is advantageous to the automotive industry. Utilizing this technique during the production of automotive body panels helps lighten total vehicle weight and improves fuel efficiency.
c. **Electronics and Telecommunications**
Both sectors of the economy make substantial use of microwave joining. Microwave joining provides a controlled and effective alternative for attaching delicate electronic components like microchips and circuit boards.
d. **Medical Devices and Biomedical Engineering**
The medical industries rely on microwave joining to form sterile and reliable connections in medical implants and equipment. The safety of patients is improved and the integrity of medical equipment is ensured by this technology.
e. **Renewable Energy**
Microwave joining is essential in the production of solar panels and wind turbine parts in the field of renewable energy. The technique makes it easier to assemble the complex components that go into these energy-efficient devices.

f. **Packaging Industry**
Materials used in food and beverage packaging, pharmaceutical blister packs, and other consumer goods are bonded by microwave joining. High efficiency is ensured in packaging operations by the rapid bonding process.
g. **Manufacturing of Electrical and Electronic Devices**
Electrical components like sensors and connections are produced via microwave joining. The dependability and performance of these crucial components are guaranteed by the procedure's precise and regulated bonding.
h. **Research and Development**
In research and development labs, microwave joining is also used to bond experimental materials and prototypes. Researchers can investigate innovative material combinations and their applications thanks to its adaptability.
i. **Industry of Textiles**
Microwave joining is used in the textile sector to join fabrics and produce specialized textiles with improved qualities. High-performance fabrics are made possible by this technology and are utilized for a variety of purposes, such as athletic gear and protective equipment.
j. **Biotechnology and Pharmaceuticals**
Materials for lab equipment and medical test devices are joined using microwave joining, which has applications in these fields of study.

The adaptability, speed, and accuracy of microwave joining have prompted its adoption in a wide range of sectors, from aerospace and automotive to electronics, healthcare, and renewable energy. The range of microwave joining applications is anticipated to broaden as technology develops, spurring innovation and efficiency in the production and research processes (Kaushal et al. 2019c, 2022).

5.2.2 Advantages of Microwave Joining Process

The use of microwave joining is favored for many industrial applications because it has a number of significant benefits over conventional bonding methods. Some of the main benefits are as follows:

a. **Enhanced Swiftness and Effectiveness**
The amazing shortening of bonding time is one of the microwave joining's most important benefits. Faster material fusion is made possible by the targeted and rapid heating process, which also results in shorter manufacturing cycles and higher production efficiency.
b. **Energy Efficiency**
The bonding technique of microwave joining is energy-efficient. Compared to bulk heating methods, it minimizes energy waste by heating only the intended locations. This lowers operating expenses while simultaneously promoting environmental sustainability.
c. **Minimal Material Distortion and Residual Tensions**
Microwave joining produces a smaller heat-affected zone than conventional bonding techniques, which might result in significant material distortion and

residual tensions. As a result, the linked materials experience less distortion and less stress, preserving the overall structural integrity.

d. **Material Compatibility**
Microwave joining is remarkably versatile and can bond a variety of materials, independent of their makeup. Microwave joining may successfully fuse many materials together, whether they are metals, ceramics, polymers, or composites, broadening its applicability across several industries.

e. **Reduced Heat-Affected Zone**
The localized heating that occurs during microwave joining reduces the heat-affected zone (HAZ) that surrounds the junction. When working with delicate materials or components that are sensitive to temperature changes, this lowered HAZ is essential.

f. **Clean and Environmentally Friendly Process**
Microwave joining does not require the use of fluxes, filler metals, or adhesives, making it a clean and environmentally friendly process. As a result, it produces less waste or emissions, making it a cleaner and greener bonding technique.

g. **High Bond Strength**
The microwave-assisted bonds are robust and long-lasting. Due to the integrity and dependability of the linked materials being guaranteed, they can be used in essential applications.

h. **Automation Potential**
Automated manufacturing processes can easily incorporate microwave joining. It is suitable for high-volume production due to the rapid bonding time and few post-processing processes.

i. **Compatible with Non-Destructive Testing (NDT)**
Microwave joining complements non-destructive testing techniques by producing few flaws and distortions. This enables quick and precise quality control checks.

j. **Precision Control**
The heating process may be precisely managed with microwave joining, leading to homogeneous bonding and repeatable outcomes. Applications that call for precise specifications benefit most from this level of control.

The benefits of microwave joining, such as improved speed, energy efficiency, low distortion, material diversity, and high bond strength, have positioned it as a cutting-edge technology that is revolutionizing the bonding of materials in numerous industries (Singh et al. 2019). It appeals to manufacturers looking for the best performance and sustainability because of its efficiency and precision.

5.2.3 Challenges and Limitations of Microwave Joining

Microwave joining has many benefits; it also has several drawbacks and restrictions that should be taken into account when using it. The following are some of the main difficulties and restrictions with microwave joining:

a. **Material Selection**
Materials should have high dielectric loss tangent values and be responsive to microwave energy for microwave joining to be most successful. As a result, only a small number of materials may be bonded directly using microwaves. When joining materials with low dielectric loss tangent values, coupling agents or susceptors may be necessary to increase energy absorption.

b. **Limitations on Thickness**
At higher frequencies, the penetration depth of microwave energy is restricted. Due to this, it may be difficult to obtain homogeneous heating in thick or large materials. Materials with considerable thicknesses may require special considerations and adaptations.

c. **Temperature Control**
To prevent overheating or underheating during the joining process, precise temperature control is essential. It can be difficult to maintain a constant and controlled temperature throughout the joining process due to the microwaves' quick heating tendency.

d. **Uneven Heating**
During microwave joining, uneven heating can be caused by nonuniform material qualities, forms, or moisture content. Localized hot or cool regions might cause unequal bonding and possible flaws.

e. **Safety Issues**
If microwave radiation is not adequately handled and controlled, it can be detrimental to human health. Operators and staff must be protected with sufficient safety measures, such as appropriate shielding, interlock systems, and personal protective equipment.

f. **Equipment Complexity**
When compared to traditional welding equipment, microwave joining equipment might be more expensive and sophisticated. The system becomes more complex and expensive as a result of the requirement for precise power and frequency regulation, as well as temperature monitoring.

g. **Considerations for Additions**
Using coupling agents, susceptors, or other additions to improve energy absorption can present extra difficulties including material compatibility and residue development.

h. **Sample Size and Geometry**
Size and shape of the samples being linked can have an impact on the effectiveness and uniformity of the heating process. To ensure homogeneous heating, irregular or complex shapes may need careful design and optimization.

i. **Scalability**
Improving the uniformity, effectiveness, and equipment scalability of microwave joining for industrial production may offer difficulties.

j. **Research and Development**
Although microwave joining has the potential to be useful, more work is still required to address the issues and improve the procedure for a wider variety of materials and applications.

Despite these obstacles and constraints, continual developments in microwave technology and process improvement are expanding the possible applications of microwave joining and providing effective and efficient substitutes for conventional joining techniques (Kaushal et al. 2018 b, 2020).

5.2.4 Factors Affecting Microwave Joining Process

The method of bonded materials together with microwave energy is known as microwave joining, microwave welding, or microwave bonding. The variables that may affect the effectiveness and success of microwave joining are discussed in the following sections.

5.2.4.1 Material Properties

The process of microwave joining is significantly influenced by the qualities of the materials. Due to their distinct dielectric characteristics, different materials react to microwave energy in different ways. Electrical characteristics known as dielectric qualities govern a material's response to an electric field. The permittivity and the loss tangent are two essential characteristics that define them. Higher loss tangent materials absorb microwaves more effectively and produce more heat during the joining process. Because they quickly absorb microwave energy and produce heat at the bonding contact, materials with greater permittivity and loss tangent values, including ceramics and specific polymers, tend to be more suitable for microwave joining (Kaushal et al. 2018 c). Conversely, to increase microwave absorption and aid the connecting process, materials with lower permittivity and loss tangent values, including metals and some insulators, may need additional measures.

5.2.4.2 Moisture Content

The microwave joining process is substantially impacted by moisture content. Moisture in the materials being connected can affect the process as a whole in both good and bad ways. To provide successful and dependable results, moisture content must be taken into account during microwave joining.

Some of the positive effects of moisture contents are as follows:

a. **Enhanced Microwave Absorption**
   Compared to many solid materials, moisture has better dielectric characteristics. As a result, materials with more moisture can absorb microwave energy more effectively, causing the joining process to heat up more quickly and effectively.
b. **Improved Energy Coupling**
   In some materials, such as wood and some polymers, moisture serves as a channel for microwave energy coupling. More efficient energy transmission may be facilitated by this improved coupling, leading to faster and more even heating.

There are some negative effects associated with moisture content.

a. **Uneven Heating**
The distribution of moisture within the material may not be consistent, which could cause uneven heating during the microwave joining process. Localized hotspots or cool spots can be brought on by variations in moisture content, which can lead to uneven bonding and associated flaws.

b. **Steam Generation**
During the heating process, excessive moisture content may cause steam to be produced. The accumulation of steam within the material may result in internal pressure, which could weaken the bond or cause delamination.

c. **Material Degradation**
Materials may be degraded, subjected to excessive moisture and microwave energy. Materials that are sensitive to moisture, such as specific electronic components or organic materials, may experience undesirable chemical reactions or changes in their physical characteristics.

5.2.4.3 Sample Geometry

The size and form of the materials being linked affect how the microwave energy is distributed and how quickly they heat up. To guarantee equal heating, the microwave setting may need to be adjusted for complex forms. The microwave joining process is substantially impacted by sample geometry. The joining process's effectiveness and quality are directly influenced by the size, shape, and distribution of the materials being joined. This relationship also affects how microwave energy is distributed and absorbed. It is essential to take into account the sample geometry and its impact on energy distribution, penetration depth, and heating rate in order to optimize the microwave joining process. Fixtures and holders that are well-designed can help ensure good contact between the materials and promote more even heating. To find the best configuration for a given joining application, experimentation and testing with various sample shapes may be required. Understanding and resolving the implications of sample geometry can improve bonding outcomes and boost the microwave joining process's overall effectiveness.

5.2.4.3.1 Pressure and Time

Sample shape has a big impact on the microwave joining process, along with pressure and time. The joining process's effectiveness and quality are directly influenced by the size, shape, and distribution of the materials being joined. This relationship also affects how microwave energy is distributed and absorbed (Pal et al. 2022).

5.2.4.4 Additives

Additives can have a substantial impact on the microwave joining process by changing the material's characteristics and affecting how it responds to microwave energy. In order to improve certain properties or speed up the bonding process, additives are compounds that are added to the materials being joined. To get the required outcomes, it is important to precisely optimize the type and concentration of additives employed during microwave joining. It is crucial to provide bonding needs, heating efficiency, and material compatibility with careful consideration. However, their application should be based on a thorough understanding of the materials and how they interact with microwave energy. Additives can provide substantial benefits in improving the microwave joining process. The bonding quality, processing speed, and overall effectiveness of microwave joining can all be improved with carefully chosen and managed additives.

5.2.4.5 Microwave Power and Distribution

The microwave joining process is greatly impacted by the microwave power and distribution. For successful and dependable bonding of the materials being connected, proper microwave power regulation and distribution are crucial. Careful consideration must be given to choosing the proper power level and guaranteeing homogeneous power distribution in order to optimize the microwave joining process. The sample size, required heating rate, and material qualities should all be taken into account when choosing the power level. For reliable and high-quality bonding results, microwave equipment with accurate power management and distribution capabilities is essential. To further guarantee that the desired temperature is attained and maintained throughout the joining process, process monitoring and temperature measurement techniques should be used. Strong and long-lasting connections between the materials are produced by the microwave joining method when microwave power and distribution are properly managed.

5.3 FRICTION STIR WELDING

FSW is often regarded as a green welding process since it uses less energy, generates less waste, and makes it simple to recycle the attached components. SW, a green technology and novel joining process was introduced by The Welding Institute (TWI), United Kingdom, in 1991. A non-consumable rotating tool generates heat and pressure and causes the metal to soften and bond together to join the base material (Sethi et al. 2021a,b, 2022; Das et al. 2023). FSW was initially created to join a variety of materials, such as aluminum alloys, copper alloy, magnesium alloy, composite material, polymer, titanium, and steel. FSW has several benefits over conventional welding methods, including the following:

a. **Environment Friendly**
   FSW is an eco-friendly welding process that does not entail melting the metal being welded, which results in lower distortion. The outcome can be a more exact and accurate joint since there is little shrinkage or distortion during the welding process.
b. **Enhanced Mechanical Properties**
   When compared to traditional welding methods, FSW can create welds with better mechanical characteristics. FSW joints often lack defects like porosity, cracking, or solidification-related defects and have a fine-grained microstructure.

c. **Higher Productivity**
   Faster than traditional welding methods, FSW is a welding procedure. According to reports, FSW can weld materials up to ten times faster than conventional welding methods, which can result in considerable financial savings.
d. **Lower Operating Costs**
   Compared to conventional welding methods, FSW uses fewer consumables like filler wire and gas shielding. Additionally, since there are no fumes or smoke produced during the process, expensive ventilation systems are not as necessary.
e. **Welding of Traditionally Unweldable Materials**
   FSW can join materials such as high-strength aluminum alloys, copper, magnesium, and titanium that are typically regarded as unweldable.

FSW has found many applications in different sectors like aerospace, automotive, shipbuilding, and internal structure of aircraft. FSW is utilized in the construction of lightweight automobiles like the Tesla Model S and the Airbus A380, as well as for joining aluminum panels on the International Space Station, among other things. To better understand the properties of FSW and to tailor the welding settings for various materials and applications, numerous research investigations have been carried out.

5.3.1 Different Stages of FSW

FSW is a multistage joining processes joint same or different metal pieces. The typical steps of the FSW process are as follows:

a. **Preparation**
   In the initial stage of FSW, the metal components that will be connected are cleaned and firmly clamped in the fixture. The fixture guarantees that the components are held firmly in place and avoids any unwanted movement during the welding process.
b. **Insertion**
   The joint line between the two metal sections is entered in the second stage using a rotating tool with a pin that has been specially created for the job. Heat and pressure are produced as the pin rotates and moves along the joint line.
c. **Stirring**
   Heat is produced due to the movement of the tool along the joint line and interaction between the tool pin and the workpiece. The metal becomes softer and stirs behind the tool as a result of the heat, forming a solid-state bond.
d. **Transverse Movement**
   The movement of tool in lateral position across the line of joint comes in the fourth stage. This motion aids in establishing a consistent weld by distributing heat and pressure equally across the joint.

e. **Withdrawal**
   The weld is allowed to cool naturally during the last stage before the tool is removed from the joint. After the weld has cooled, it is examined for flaws and given any necessary post-weld treatments.

5.3.2 Factors Affecting FSW

In order to form a solid-state bond between two pieces of metal, FSW is a multi-stage, complicated operation. To create high-quality welds, the procedure needs specialized tools and knowledgeable operators. The quality and performance of FSW are influenced by several process parameters, including the following:

a. **Tool Rotation Speed**
   One of the most important process variables in FSW is the tool rotation speed. The metal becomes softer and is able to flow around the tool due to the generation of frictional heat by the rotating tool. The type of metal being welded and its thickness affect the ideal rotation speed.
b. **Tool Traverse Speed**
   The rate at which the tool traverses along the line of joint is referred to as tool traverse speed (TTS). The ideal traversal speed depends on the welding conditions, the welding material, and the tool design.
c. **Tool Tilt Angle**
   The angle formed by the tool axis and the welding area is referred to as the tool tilt angle (TTA). The quality of the weld and the flow of material around the tool can both be impacted by the tilt angle.
d. **Axial Force**
   The force used to keep the tool's penetration depth accurate is known as the axial force. The material being welded and the tool design affect the ideal axial force.
e. **Welding Speed**
   The tool's motion along the joint line is referred to as the welding speed. The amount of heat applied to the metal and the weld quality can both be impacted by welding speed.
f. **Shoulder Diameter**
   The tool's shoulder diameter is known as the shoulder diameter. During the welding process, the shoulder diameter may have an impact on the heat input and pressure distribution. Maximum heat (85–95%) is generated by the interaction of the tool shoulder and workpiece.
g. **Pin Profile**
   The tool pin's shape is referred to as the pin profile. The pin profile may have an impact on the weld's quality and material flow (Kumar et al. 2022c).

5.4 THE EFFECT OF TOOL ROTATIONAL SPEED ON FSW SPECIMEN

The tool rotational speed (TRS) plays a significant role to effect on the efficacy and quality of the FSW process. The welding material, joint geometry, and desired weld quality influence the optimal rotating speed for FSW. It is essential to select a rotation speed that optimizes the formation of heat, the movement of material, and tool wear to produce welds of the most excellent quality with the fewest potential faults. In FSW, the tool rotational speed is one of the most crucial process variables and significantly affects the properties of the welded specimens. The following are a few impacts of rotational tool speed on FSW specimens.

5.4.1 Heat Generation

The heat produced during welding also depends on the tool's rotational speed due to significant frictional interaction between the stir tool and the workpiece. Increasing the tool's rotating speed accelerates the heat generation rate. As a result of this enhanced heat generation, weld temperatures may increase, lowering welding forces and promoting material flow. However, overheating a material due to extremely high rotational speeds can lead to undesired material microstructure and flaws. Using the Artificial Neural Network (ANN) technique, Kumar et al. (2023d) discussed the impact of heat generation on mechanical characteristics during different ranges of TRS from 600 rpm to 1800 rpm in the FSW process. It was found that as TRS is increased, the spindle torque, X-force, and force acting Z-force all decrease. The mechanical properties of welded samples revealed an increasing of TRS up to 1400 rpm and subsequently a declining trend for 1800 rpm. TRS increases from 600 to 1800 rpm, and frictional heat production and grain size increase at the nugget zone (NZ). Kaushik and Kumar (2023) investigated the heat generation during welding of dissimilar materials (aluminum and steel). A unique method is used to quantify the discontinuities (steel debris and voids) in the weld stir zone. Additionally, parametric research using a full factorial method and three tool rotational and traverse speed levels was carried out, and a heat input factor was derived using theoretical process parameter modeling. The degree of discontinuities affected the soundness and strength of the weld joints, according to the results. Maximum joint strength was produced at a low rotational speed of 386 revolutions per minute and a moderate traverse speed of 140 mm per minute with few discontinuities. Kumar et al. (2023a) joined 2050Al-Li alloy at varying TRS (700 rpm to 1600 rpm) with TTS of 10 mm/s and TTA of 2^0. The result revealed that increasing TRS reduces the weldment surface's half-circle space on the weld bead and grain size. However, the heat input during FSW increases as TRS grows. At higher TRS (1600 rpm), the joint efficiency and tensile strength are at their highest levels. Kumar et al. (2023b) observed with increasing TRS, the heat generation per unit length increases due to interaction between the tool and workpiece as shown in Figure 5.1.

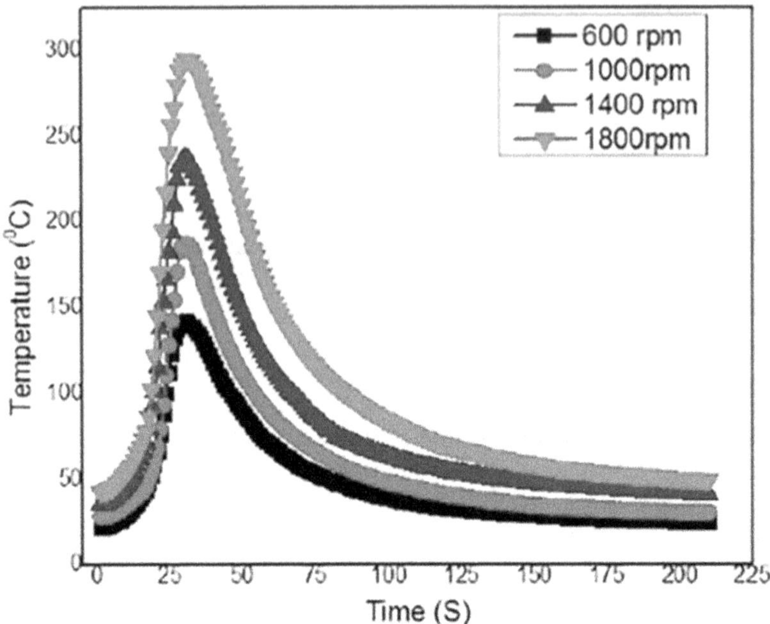

FIGURE 5.1 Effect of varying TRS on heat generation during FSW.

Source: Kumar et al. (2023).

5.4.2 Weld Quality

The tool's rotating speed impacts the weld's quality. The material may not soften sufficiently to establish a suitable connection if the tool's spinning speed is too slow, leading to a weak or insufficient weld. The material may overheat if the tool's rotational speed is too high, leading to flaws like voids or cracks. Khan et al. (2021) analyzed the effect of varying TRS (700 to 1120 rpm) on the weld quality of friction stir welded (FSWed) joints of 6061 aluminum alloy. Results showed that most FSWed samples produced a bell-shaped bead. The stir tool's shoulder diameter and pin diameter were approximately equivalent to the top and bottom surfaces of the weld bead, respectively. Wang et al. (2020) examined the impact of changing TRS from 300 rpm to 600 rpm at a constant TTS of 100 mm/min on the mechanical characteristics of FSWed steel plates (1.86 mm thick, UNS S32205). Due to deficient heat input, the partial penetration defect developed at 300 rpm, and the severe tool sticking caused a groove-like defect to develop at 600 rpm. Welds without defects were produced at 350 to 500 rpm. Within the NZ and TMAZ, finer recrystallized grains were seen. As the rotational speed increased, the joint width increased, and higher strength and hardness were observed at lower TRS values. Kumar et al. (2022a) joined Al-Li alloy at different TRS and observed fine structure at higher TRS as shown in Figure 5.2.

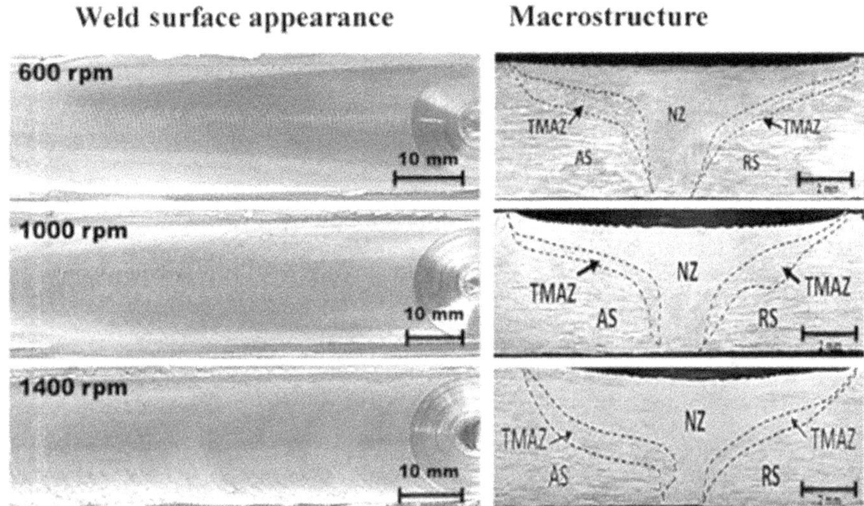

FIGURE 5.2 Weld bead surface structure and respective zone at different TRS.

Source: Kumar et al. (2022).

5.4.3 Microstructure

The tool's rotational speed impacts the welded specimens' microstructure. Because friction produces more heat at faster rotating speeds, the grain structure becomes finer and more uniform. In contrast, a coarse grain structure can be produced at lower rotational speeds. Using FSW, Mehri et al. (2023) investigated the effects of various TRS of 600 rpm to 1600 rpm on a 7075-T6 Al alloy sheet. A satisfactory welding condition has been identified from the mechanical strength at 1000 rpm and TTS of 50 mm/min. As welding rotational speed increases, more base material precipitates are dissolved in the metallic matrix, and their distribution becomes more uniform. It is understood that this phenomenon is the primary root cause of the mechanical properties decreasing at 1600 rpm. Kumar et al. (2023) joined the material at different TRS and observed lower grain size at TRS of 1600 rpm and TTS of 10 mm/s as shown in Figure 5.3, providing higher strength.

5.4.4 Mechanical Characteristics

TRS is the main parameter which affects the mechanical properties of the FSWed samples. Due to the increasing amount of heat input at more incredible rotating speeds, the weld gains strength and ductility. In contrast, excessive rotational speed can result in material displacement and deformation, reducing the mechanical characteristics of the material. Kumar et al. (2022b) revealed defect-free joints at different ranges of TRS from 600 rpm to 1400 rpm. The grain size at NZ and TMAZ increases as TRS (600–1400 rpm) increases. At 1400 rpm, the material possessed a

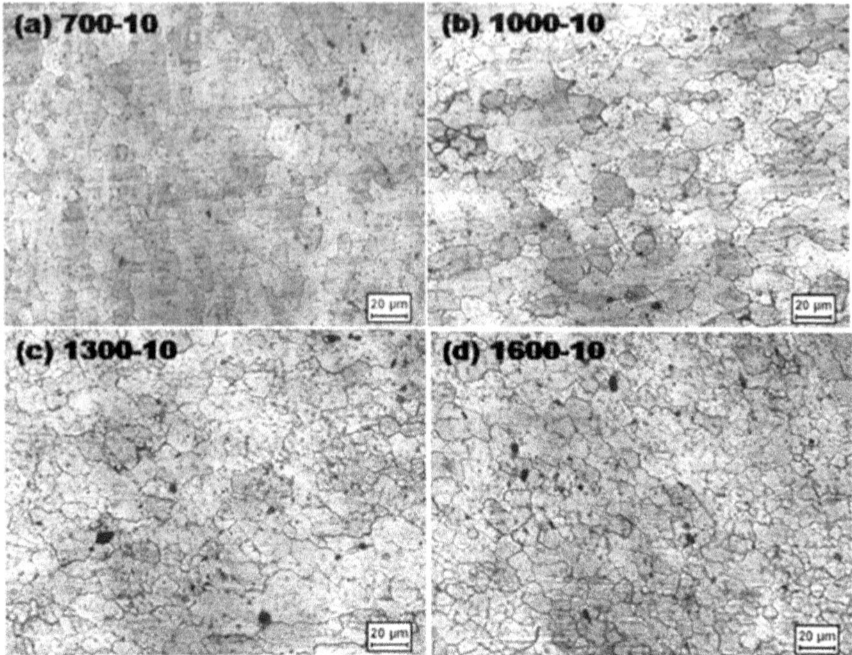

FIGURE 5.3 Effect of different TRS: (a) 700 rpm, (b) 1000 rpm, (c) 1300 rpm, (d) 1600 rpm on microstructure at nugget zone.

Source: Kumar et al. (2023).

high ultimate tensile strength and percent elongation. He et al. (2016) studied the effects of varying TRS on thick AA 6061-T6 FSWed plates. They noticed that as rotation speed increased, the tensile strength increased to its maximum value before decreasing as rotation speed increased further. They explained this behavior by the higher rotating speed and increased heat input, which caused the Mg_2Si precipitates to dissolve and the grain to become coarser. In a different investigation, Hou et al. (2014) noted that when the joint's rotation speed rises, so do the NZ's grain size and dislocation density. Kumar et al. (2023d) joined 2050-Al-Li alloy at different TRS and observed higher strength at higher TRS of joined samples as shown in Figure 5.4.

5.4.5 Material Flow

The amount of material flowing around a tool depends on its rotation rate. Due to more excellent frictional heating, there is an increase in the flow of material around the tool at faster rotational speeds. This may lead to a more even and steady material flow around the tool.

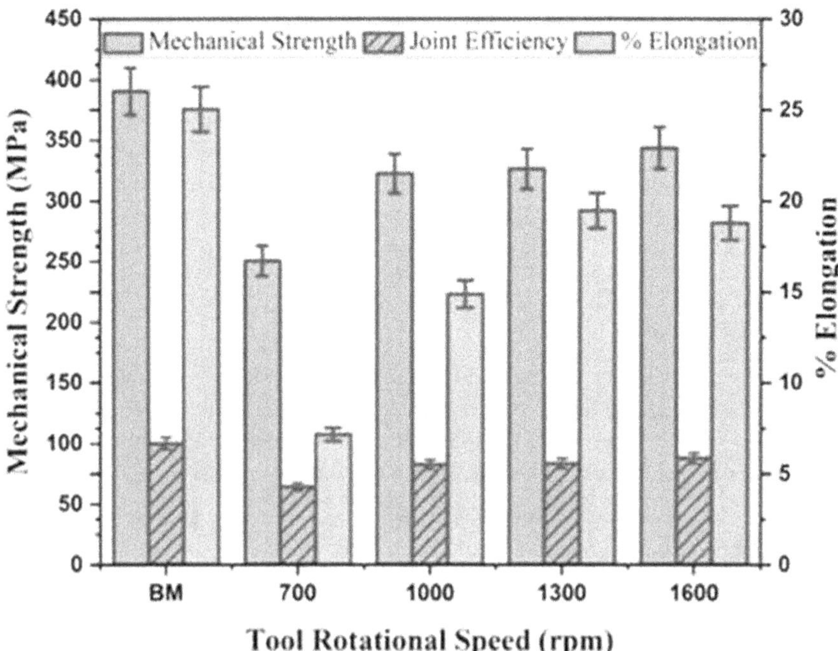

FIGURE 5.4 Effect of different TRS on mechanical strength.

Source: Kumar et al. (2023).

5.4.6 Tool Wear

The TRS affects the tool's wear and tear. High TRS may increase tool wear, shorten tool life, necessitate more frequent tool changes, and raise welding costs. The impact of TRS on the mechanical characteristics of the FSWed aluminum alloy is examined by Liu et al. (2018). Higher TRS results in more refined grains, greater material flow, stronger joints, and higher hardness. However, extremely high TRS can result in voids and tunnel flaws. Li et al. (2017) observed that increasing TRS causes the grain structure to become more refined, joint strength to increase, and fatigue characteristics to improve. The study highlights the significance of choosing the best TRS to strike a compromise between grain refinement and defect development. Zhang et al. (2015) observed that increasing TRS causes the grain structure to become more refined, the porosity to decrease, and the joint strength to increase. However, significantly higher TRS might result in fractures and tunnel flaws, and a similar finding was observed by Khalid et al. (2022).

The tool rotational speed is a critical parameter in FSW, and it must be carefully controlled to achieve high-quality welds with optimal properties. The optimal rotational speed depends on several factors, including the material being welded, the thickness of the metal, and the desired properties of the final weld.

5.5 THE EFFECT OF THE TOOL TRAVERSE SPEED ON FSWED SPECIMEN

The TTS is an important parameter in FSW, and it has a significant effect on the properties of the welded specimens. The rate at which the tool moves along the joint while welding is referred to as the TTS. Typically, it is expressed in inches per minute (in/min) or millimeters per minute (mm/min). The quality and characteristics of the final weld can all be impacted by the TTS during the welding process. Following are some instances of the way that TTS affects FSW specimens.

5.5.1 Weld Quality

The TTS affects the quality of the weld. If the traverse speed is too low, the material may not soften enough to form a proper bond, resulting in a weak or incomplete weld. If the traverse speed is too high, the material may not have enough time to properly flow around the tool, resulting in defects such as voids or cracks.

5.5.2 Microstructure

The TTS affects the microstructure of the welded specimens. At higher TTS, the heat input to the material is lower, resulting in a coarse grain structure. Conversely, lower TTS can result in a finer and more uniform grain structure. The impact of TTS on the microstructure of FSWed Al-Li alloy was investigated by Kumar et al. (2020). The NZ had equiaxed grains because of higher heat output, and the TMAZ had elongated grains due to reduced material deformation. The average grain size in NZ reduces from 19.84 m to 14.86 m when TTS is increased from 1 mm/s to 4 mm/s. The effect of WS on the microstructure of an FSWed aluminum joint was studied by Sakthivel et al. (2019). Lower WS is found to have more homogenous grains than higher welding speeds. The impact of TTS on the microstructural characteristics of FSW joints constructed of 6005AT6 aluminum alloy was examined by Dong et al. (2013). The NZ has a fine, equiaxed grain structure mainly composed of the Al solid solution. While the grain structure of HAZ is coarse equiaxed of TMAZ is distinguished by elongated grains with a high number of dislocations.

5.5.3 Mechanical Properties

TTS affects the mechanical properties of the FSWed specimens. The optimal traverse speed results in a weld with high strength and ductility. However, if the traverse speed is too high or too low, the mechanical properties of the FSWed specimens may be compromised. Kumar et al. (2020) joined third generation of Al-Li alloys at varying TTS and observed that with increasing TTS, the tensile strength increased as shown in Figure 5.5.

5.5.4 Heat Input

The heat produced during welding depends on the TTS. While slowing down can increase heat input, a more incredible traversal speed can result in a reduction

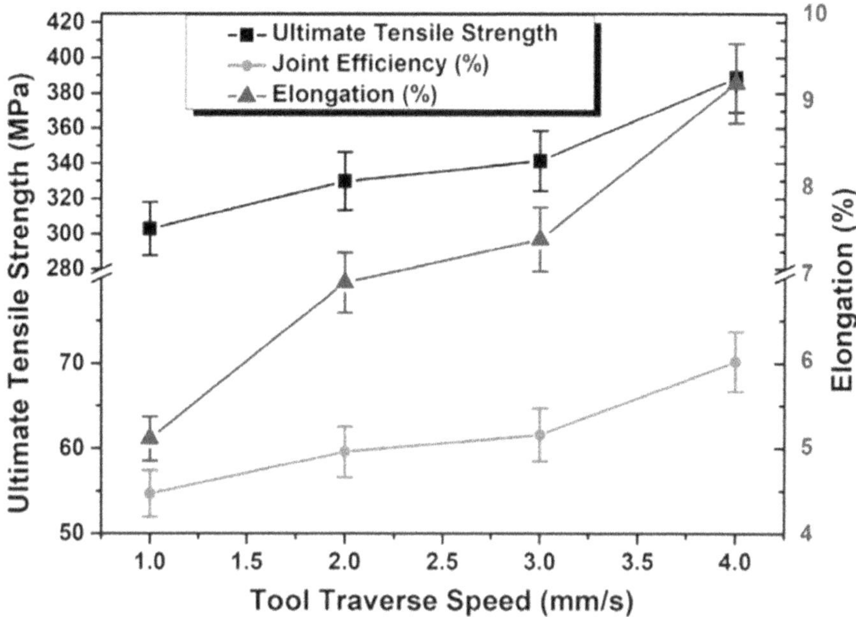

FIGURE 5.5 Effect of varying TTS on friction stir welded specimen.

Source: Kumar et al. (2020).

(Kumar et al. 2023c). The amount of heat applied influences the plasticized material's softening and flow properties, affecting the weld quality and leading to flaws such as voids or insufficient bonding.

5.5.5 Size of the Stir Zone

The TTS influences the size of the material's stirred or plasticized zone. While a lower speed can result in a larger and deeper stir zone, a higher traverse speed often creates a shallower and narrower stir zone. The strength and fatigue resistance of the weld are mechanical qualities that are influenced by the size of the stir zone.

5.5.6 Material Flow

During FSW, the traverse speed impacts the material flow. A lower speed enables more material displacement, whereas a faster speed may result in less movement of plasticized material. The weld's base metal, oxide layers, and other contaminants are distributed differently depending on the material flow. The weld's mechanical qualities may be impacted by how it affects the grain structure and texture.

Overall, the TTS is a critical parameter in FSW, and it must be carefully controlled to achieve high-quality welds with optimal properties. The optimal traverse speed depends on several factors, including the material being welded, the thickness of the metal, and the desired properties of the final weld.

5.6 EFFECT OF THE TOOL TILT ANGLE ON FSWED SPECIMEN

The TTA significantly impacts the parameters of the welded specimens, another crucial process parameter in FSW. In order to transport the deformed material efficiently to the tool's backside, it is necessary to choose an appropriate tilt angle. The TTA has a significant effect on the formation of heat and weld defects. As the TTA decreases, plastic material flow is restricted. In addition, as the tilt angle increases, a flash is produced on the retreating side of the nugget zone, causing cavities to form. The following are a few impacts of TTA on FSW specimens:

5.6.1 Welding Quality

The TTA affects the quality of joint. Higher or lower TTA could prevent the material from correctly flowing around the tool, leading to flaws like voids or fissures. The ideal tilt angle might change based on the metal thickness and the substance being welded.

5.6.2 Microstructure

The TTA impacts the welded specimens' microstructure. The generation of friction during joining of material is decreased at higher tilt degrees, resulting in a coarse grain structure. On the other hand, a finer and more uniform grain structure may be produced by lower tilt angles.

5.6.3 Mechanical Characteristics

The TTA influences the welded specimens' mechanical characteristics. A weld with the ideal tilt angle has high strength and ductility. The mechanical qualities of the welded specimens could be compromised if the tilt angle is too high or too low. The impact of TTA on the microstructural characteristics of FSW of AA 6061-T6 aluminum alloy was examined in a study by Banik et al. (2018). According to the analysis, a joint's tensile characteristics rise with the angle of tilt increases. Acharya et al. (2021) looked into how TTA affected the mechanical characteristics of AA6092/17.5 SiCp-T6 plates. It has been noted that as TTA rises from 1° to 2°, the mechanical characteristic of the joint increases, and as TTA rises from 2° to 3°, it decreases.

5.6.4 Heat Input

The TTA influences the heat input to the base material being joined. The temperature rise in the welded zone is reduced at higher tilt angles because the heat input is smaller. It may impact the mechanical characteristics of the welded material. In FSW, the TTA is a significant parameter that must be carefully regulated to produce welds with the best possible characteristics. Several elements, such as the welding material, the thickness of the metal, and the desired weld characteristics, determine the ideal tilt angle. Kumar et al. (2023b) found that joint at 2° of TTA produced higher tensile strength and joint at different TTA is depicted in Figure 5.6.

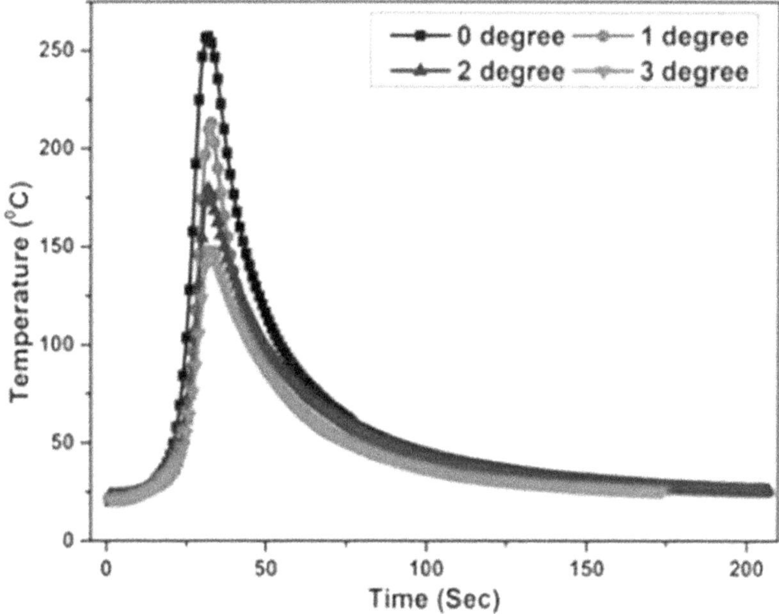

FIGURE 5.6 Effect of different TTA on welded specimen.

Source: Kumar et al. (2023).

5.7 EFFECT OF THE TOOL SHOULDER ON FSWED SPECIMEN

The FSW tool shoulder is a crucial component and significantly affects the characteristics of the welded specimens. The following are a few outcomes of the tool shoulder on FSW specimens:

5.7.1 Heat Input

The quantity of heat input to the welded material is predisposed by the tool shoulder. Due to the increased contact area between the tool and the material, larger tool shoulders often generate more heat. This may impact the microstructure and mechanical characteristics of the welded specimens.

5.7.2 Material Flow

The shoulder of the tool has an impact on how the material moves around the tool. The material flow around the tool may be more even and steady with a more extensive tool shoulder diameter. The tool shoulder's size, on the other hand, might result in excessive friction and material displacement, which can result in flaws like voids or cracks.

5.7.3 Weld Quality

Tool shoulders have an impact on the weld's quality. A better weld can be produced because there is more material in contact with the tool when its shoulder is more significant. However, if the tool shoulder is too big, it may lead to excessive material displacement and distortion, resulting in flaws in the welded specimens.

5.7.4 Mechanical Characteristics

The mechanical characteristics of the welded specimens are influenced by the tool shoulder. A weld with more strength and ductility may be produced by a more prominent shoulder diameter tool. However, if the tool's shoulder is too big, it may lead to excessive material displacement and distortion, impairing its mechanical qualities.

In FSW, the tool shoulder is a crucial variable that must be selected appropriately to produce welds with the best possible characteristics. Several elements, such as the welding substance, the metal thickness, and the desired weld characteristics, influence the ideal tool shoulder diameter.

5.8 EFFECT OF TOOL PIN PROFILE ON FSWED SPECIMEN

A different type of tool pin profile is used to join the material having different aspects. Some material produced prominent mixing of plasticized material, while others have higher deformation. Various authors have joined the material by using different tools. About 15% of the heat produced comes from the tool pin, which also affects how plasticized material mixes and flows from the advancing side (AS) to the retreating side (RS). Kumar et al. (2021) observed that hybrid pin profile (coupling of taper threaded and triangular tool pin) produced higher strength as compared to taper threaded and triangular tool at same TRS and TTS. According to Leon et al. (2020), the hexagonal tool produces more heat than the triangular tool pin at a higher shoulder cone angle (4.5°) and a lower tool pin cone angle (5°) because of the tool's many sides, which control the flow rate of plasticized material around the tool and under the shoulder. Ugender et al. (2018) found that compared to different tool pin profiles such as cylindrical, threaded conical, and conical profiles, the taper threaded profile tool pin offers greater mechanical qualities (tensile strength and micro-hardness). Regarding the mechanical characteristics and joint efficiency of FSWed AA5083 alloy plates at lower TTS, the triangular pin profile is preferable to the threaded circular profile pin (Birol et al. 2013). According to Elangovan et al. (2007), the square profile tool pin has better tensile qualities than other tool pin profiles at an axial force of 7KN. Palanivel et al. (2012) made comparable observations while utilizing a different profile tool pin and an axial force of 1.5 tonnes.

5.9 EFFECT OF PLUNGE DEPTH ON FSWED SPECIMEN

The tool's plunge depth, or the depth to which it is inserted into the workpiece, is a crucial factor that influences the effectiveness and caliber of the welding process.

With increased plunge depth, mechanical prosperities decreased, and faulty welds occurred, according to Devanathan et al. (2013). According to Kandasamy et al. (2012), a reduced dive depth produced weld flaws since the deformed material did not stay in the weld cavity when welding. Additionally, greater plunging depth caused the weld zone to thin out due to material loss from flash. Additionally, they concluded that temperature rose gradually as diving depth decreased as TTA rose from 2° to 3°. The impact of plunge depth on FSW is as follows:

5.9.1 Heat Source

The amount of heat produced during welding is influenced by the tool's plunge depth. The heat input rises as the plunge depth deepens because more material comes into contact with the tool. This may cause a more significant heat-affected zone and more softening. On the other hand, a shallower plunge depth produces a smaller heat-affected zone and less heat intake.

5.9.2 Material Flow

During the welding process, the tool's plunge depth impacts the material flow and mixing. A weld can be smoother and more uniform if the plunge depth is greater because more material is displaced and mixed during welding. However, an intense dive might result in surface flaws and material extrusion. Less material is moved and mixed when the plunge depth is smaller, which might lead to a less complete and uniform weld.

5.9.3 Weld Quality

The tool's plunge depth impacts the general weld quality. Choosing the proper plunge depth can produce a strong and dependable weld with good mechanical qualities. An inadequate plunge depth might result in flaws like voids and surface fractures.

5.10 CONCLUSION

FSW, a green energy-based welding technique, has technical and significant environmental advantages. FSW decreases the emission of harmful compounds into the atmosphere and removes the need for consumable filler materials, which improves air quality and has a minimal negative impact on human health and the environment. FSW also requires less energy input, which results in decreased energy use and carbon emissions. The approach also produces less waste by removing the requirement for additional materials frequently utilized in conventional welding procedures. FSW may be used on various materials, creating high-quality welds with outstanding mechanical qualities. Overall, an FSW supports a more environmentally friendly and sustainable approach to metal joining operations and represents a greener alternative to traditional fusion welding techniques.

REFERENCES

Acharya, U., Roy, B. S., & Saha, S. C. (2021). On the role of tool tilt angle on friction stir welding of aluminum matrix composites. *Silicon*, 13(1), 79–89. https://doi.org/10.1007/s12633-020-00405-5.

Banik, A., Roy, B. S., Barma, J. D., & Saha, S. C. (2018). An experimental investigation of torque and force generation for varying tool tilt angles and their effects on microstructure and mechanical properties: Friction stir welding of AA 6061-T6. *Journal of Manufacturing Processes*, 31, 395–404. https://doi.org/10.1016/j.jmapro.2017.11.030

Birol Y., & Kasman S. (2013). Effect of welding parameters on microstructure and mechanical properties of friction stir welded EN AW 5083 H111 plates. *Materials Science and Technology*, 29(11), 1354–1362. https://doi.org/10.1179/1743284713Y.0000000280.

Das, R., Kumar, S., Katiyar, J. K., Choudhury, S., & Roy, B. S. (2023). State-of-the-art on microstructural, mechanical and tribological properties of friction stir processed aluminium 2xxx series alloy. *Proceedings of the Institution of Mechanical Engineers, Part E: Journal of Process Mechanical Engineering*, 09544089231215223.

Devanathan, C., & Babu, A. S. (2013). Effect of plunge depth on Friction Stir Welding of Al 6063. International Conference on Advanced Manufacturing and Automation (INCAMA-2013).

Dong, P., Li, H., Sun, D., Gong, W., & Liu, J. (2013). Effects of welding speed on the microstructure and hardness in friction stir welding joints of 6005A-T6 aluminum alloy. *Materials & Design*, 45, 524–531. https://doi.org/10.1016/j.matdes.2012.09.040.

Elangovan K., Balasubramanian V., & Valliappan M. (2007). Influences of tool pin profile and axial force on the formation of friction stir processing zone in AA6061 aluminum alloy. *The International Journal of Advanced Manufacturing Technology*, 38(3–4), 285–295. https://doi.org/10.1007/s00170-007-1100-2

He, J., Ling, Z., & Li, H. (2016). Effect of tool rotational speed on residual stress, microstructure, and tensile properties of friction stir welded 6061-T6 aluminum alloy thick plate. *The International Journal of Advanced Manufacturing Technology*, 84, 1953–1961. https://doi.org/10.1007/s00170-015-7859-7

Hou, J. C., Liu, H. J., & Zhao, Y. Q. (2014). Influences of rotation speed on microstructures and mechanical properties of 6061-T6 aluminum alloy joints fabricated by self-reacting friction stir welding tool. *The International Journal of Advanced Manufacturing Technology*, *73*, 1073–1079. https://doi.org/10.1007/s00170-014-5857-9

Kandasamy, K., Kailas, S. V., & Srivatsan, T. S. (2012). the extrinsic influence of tool plunge depth on friction stir welding of an aluminum alloy. In *Advanced Materials Research* (Vol. 410, pp. 206–215). Trans Tech Publications Ltd. https://doi.org/10.4028/www.scientific.net/AMR.410.206

Kaushal, S. (2022). Microstructure and tribological characterization of composite castings developed through in-situ microwave hybrid heating. *International Journal of Metalcasting*, 16(4), 2150–2161.

Kaushal, S., Bohra, S., Gupta, D., & Jain, V. (2020). On processing and characterization of Cu–Mo-based castings through electromagnetic heating. *International Journal of Metalcasting*, 15(2), 530–537, DOI:10.1007/s40962-020-00481-8

Kaushal, S., Gupta, D., & Bhowmick, H. (2017). On microstructure and wear behavior of microwave processed composite clad. *Journal of Tribology*, 139(6), 061602.

Kaushal, S., Gupta, D., & Bhowmick, H. (2018a). An approach for functionally graded cladding of composite material on austenitic stainless steel substrate through microwave heating. *Journal of Composite Materials*, 52(3), 301–312.

Kaushal, S ., Gupta, D., & Bhowmick, H. (2018b). On surface modification of austenitic stainless steel using microwave processed Ni/Cr3C2 composite cladding. *Surface Engineering*, 34(11), 809–817.

Kaushal, S., Gupta, D., & Bhowmick, H. (2018c). On processing of Ni-WC based functionally graded composite clads through microwave heating. *Materials and Manufacturing processes*, 33(8), 822–828.

Kaushal, S., Gupta, D., & Bhowmick, H. (2019a). Corrigendum: On processing of Ni-Cr3C2 based functionally graded clads through microwave heating. *Materials Research Express*, 6, 109501, DOI: 10.1088/2053-1591/ab3747.

Kaushal, S., Gupta, D., & Bhowmick, H. (2019b). On processing and flexural behaviour of functionally graded clads developed through microwave irradiation. *Materials Research Express*, 6(7), 076405. DOI 10.1088/2053-1591/ab11ea.

Kaushal, S., Singh, D., Gupta, D., & Jain, V. (2019c). Wear resistance improvement of Austenitic 316 L steel by microwave-processed composite clads. *ASME Journal of Tribology*, 141(4), 041605. https://doi.org/10.1115/1.4042273

Kaushal, S., Gupta, D., & Bhowmick, H. (2021). Wear behavior of microwave-processed Ni-WC8Co-based functionally graded materials. *Proceedings of the Institution of Mechanical Engineers, Part L: Journal of Materials: Design and Applications*, 235(5), 1036–1045. doi:10.1177/1464420720988119

Kaushal, S., Singh, S., & Gupta, D. (2022). Processing strategy for high strength Ni-based hybrid composite clad on SS 316L steel through microwave heating. *Proceedings of the Institution of Mechanical Engineers, Part B: Journal of Engineering Manufacture*, 236(3), 190–203. doi:10.1177/09544054211021360

Kaushik, P., & Kumar, D. (2023). Heat generation and steel fragment effects on friction stir welding of aluminum alloy with steel. *Materials and Manufacturing Processes*, 1–13. https://doi.org/10.1080/10426914.2023.2187827

Khalid, E., Shunmugasamy, V. C., & Mansoor, B. (2022). Microstructure and tensile behavior of a Bobbin friction stir welded magnesium alloy. *Materials Science and Engineering: A*, 840, 142861. https://doi.org/10.1016/j.msea.2022.142861

Khan, N., Rathee, S., Srivastava, M., & Sharma, C. (2021). Effect of tool rotational speed on weld quality of friction stir welded AA6061 alloys. *Materials Today: Proceedings*, 47, 7203–7207. https://doi.org/10.1016/j.matpr.2021.07.496.

Kumar, S., Acharya, U., Sethi, D., Medhi, T., Roy, B.S., & Saha, S.C. (2020). Effect of traverse speed on microstructure and mechanical properties of friction-stir-welded third-generation Al–Li alloy. *Journal of the Brazilian Society of Mechanical Sciences and Engineering*, 42(8), 1–13. https://doi.org/10.1007/s40430-020-02509-w.

Kumar, S., Chaubey, S.K., Sethi, D., Saha, S.C., & Roy, B.S., (2021). Performance analysis of varying tool pin profile on friction stir welded 2050-T84Al-Cu-Li alloy plates. *Journal of Materials Engineering and Performance*, pp.1–12. www.researchgate.net/publication/355589422_Performance_Analysis_of_Varying_Tool_Pin_Profile_on_Friction_Stir_Welded_2050-T84Al-Cu-Li_Alloy_Plates

Kumar, S., Choudhury, S., Sethi, D., Paulraj, J., Bhargava, M., & Saha Roy, B. (2022a), Effect of process parameters on third generation of friction stir welded Al-Li alloy plates, *CIRP Journal of Manufacturing Science and Technology*, 38, 372–385. https://doi.org/10.1016/j.cirpj.2022.05.009.

Kumar, S., & Katiyar, J. K., (2023).Roles of tribology in friction stir welding and processing, in tribology in sustainable manufacturing, 162–188. DOI: 10.1201/9781003363576-9.

Kumar, S., Katiyar, J. K., Acharya, U., Saha, S. C., & Roy, B. S. (2022b). Influence of tool rotational speed on Microstructure and Mechanical Properties of Al-Li Alloy using Friction

Stir Welding. *Proceedings of the Institution of Mechanical Engineers, Part E: Journal of Process Mechanical Engineering*, 09544089221080823.

Kumar, S., Katiyar, J. K., Kesharwani, G. S., & Roy, B. S. (2023a). Microstructure, mechanical, and force-torque generation properties of friction stir welded third generation Al/Li alloy at higher traverse speed. *Materials Today Communications*, 106084.

Kumar, S., Katiyar, J. K., & Roy, B. S. (2023b). Influence of tool tilt angle on physical, thermal, and mechanical properties of friction stir welded Al-Cu-Li alloys. *Materials Today Communications*, 105348. https://doi.org/10.1016/j.mtcomm.2023.105348.

Kumar, S., Sethi, D., Choudhury, S., Das, R., Saha, S. C., & Saha Roy, B. (2022c). Effect of tool pin profiles on surface roughness of friction stir welded 2050-T84 Al-Cu-Li alloys, *Journal of Mines, Metals and Fuels*, 70(3A), 108–113. ISSN0022-2755

Kumar, S., Soni, A., Katiyar, J. K., Kumar, S., & Saha Roy, B. (2023c). Influence of tool pin profiles on waviness and natural frequency during friction stir welding of Al-Li alloys plates. *Surface Topography: Metrology and Properties,* 11, 025015. DOI 10.1088/2051-672X/acd5ea

Kumar, S., Triveni, M. K., Katiyar, J. Nath, T., & Roy, B. S. (2023d). Prediction of heat generation effect on force torque and mechanical properties at varying tool rotational speed in friction stir welding using Artificial Neural Network, *Part C: Journal of Mechanical Engineering Science*, 1–20. DOI: 10.1177/09544062231155737.

Leon, J. S. & Jaya Kumar, V. (2020). Effect of tool shoulder and pin cone angles in friction stir welding using non-circular tool pin. *Journal of Applied and Computational Mechanics*, 6(3), 554–563. DOI: 10.22055/JACM.2019.29340.1585

Li, Y., Qin, F., Liu, C., & Wu, Z. (2017). A review: Effect of friction stir welding on microstructure and mechanical properties of magnesium alloys. *Metals*, 7(12), 524. https://doi.org/10.3390/met7120524

Liu, F. J., Fu, L., & Chen, H. Y. (2018). Effect of high rotational speed on temperature distribution, microstructure evolution, and mechanical properties of friction stir welded 6061-T6 thin plate joints. *The International Journal of Advanced Manufacturing Technology*, 96, 1823–1833. https://doi.org/10.1007/s00170-018-1736-0

Mehri, A., Abdollah-zadeh, A., Entesari, S., Saeid, T., & Wang, J. T. (2023). The effects of friction stir welding on microstructure and formability of 7075-T6 sheet. *Results in Engineering*, 18, 101041. https://doi.org/10.1016/j.rineng.2023.101041.

Pal, J., Gupta, D., & Singh, T.P. (2022). Processing and characterization of SS316 based metal matrix composite casting through microwave hybrid heating. *Proceedings of the Institution of Mechanical Engineers, Part C: Journal of Mechanical Engineering Science*, 236(20), 10508–10527. doi:10.1177/09544062221104443

Palanivel, R., Koshy Mathews, P., Balakrishnan, M., Dinaharan, I., & Murugan, N. (2012). Effect of tool pin profile and axial force on tensile behavior in friction stir welding of dissimilar aluminum alloys, *Advanced Materials Research*, 415, 1140–1146. https://doi.org/10.4028/www.scientific.net/AMR.415-417.1140

Sakthivel, T., Sengar, G. S., & Mukhopadhyay, J. (2019). Effect of welding speed on microstructure and mechanical properties of friction-stir-welded aluminum. *The International Journal of Advanced Manufacturing Technology*, 43(5), 468–473. https://doi.org/10.1007/s00170-008-1727-7

Sethi, D., Acharya, U., Kumar, S., Shekhar, S., & Roy, B.S., (2021a). Effect of reinforcement particles on friction stir welded joints with scarf configuration: an approach to achieve high strength joints. *Silicon*, 1–14. https://doi.org/10.1007/s12633-021-01430-8.

Sethi, D., Acharya, U., Kumar, S., Shekhar, S., & Roy, B.S. (2022). Effect of tool rotational speed on friction stir welded AA6061-T6 scarf joint configuration, *Advanced Composites and Hybrid Materials*, https://doi.org/10.1007/s42114-022-00434-1-.

Sethi, D., Kumar, S., Shekhar, S., & Roy, B.S. (2021b). Friction stir welding of AA7075-T6/TiB2 in situ cast composites plates using scarf joint configuration. *Advances in Materials and Processing Technologies*, 1–13. https://doi.org/10.1080/2374068X.2021.1959092

Silberglitt, R., Ahmad, I., Black, W.M. *et al.* (1993). Recent developments in microwave joining. *MRS Bulletin*, 18, 47–50. www.cambridge.org/core/journals/mrs-bulletin/article/abs/recent-developments-in-microwave-joining/611D62A97223F0EA801C929E34BDB130

Singh, S., Singh, P., Gupta, D., Jain, V., Kumar, R., & Kaushal, S. (2019. Development and characterization of electromagnetic processed cast iron joint. *International Journal of Engineering Science Technologies*, 22(2), 569–577.

Ugender, S. (2018). Influence of tool pin profile and rotational speed on the formation of friction stir welding zone in AZ31 magnesium alloy, *Journal of Magnesium and Alloys*, 6(2), 205–213. https://doi.org/10.1016/j.jma.2018.05.001.

Wang, W., Hu, Y., Wu, T., Zhao, D., & Zhao, H. (2020). Effect of rotation speed on microstructure and mechanical properties of friction-stir-welded 2205 duplex stainless steel. *Advances in Materials Science and Engineering*, 1–13. https://doi.org/10.1155/2020/5176536.

Zhang, F., Su, X., Chen, Z., & Nie, Z. (2015). Effect of welding parameters on microstructure and mechanical properties of friction stir welded joints of a super high strength Al–Zn–Mg–Cu aluminum alloy., 67, 483–491. https://doi.org/10.1016/j.matdes.2014.10.055

6 Green Approach for Synthesis of Nanomaterials

Rohit Lade

6.1 INTRODUCTION

Nanotechnology is an ever-evolving field with continuous new advancements. This has generated an immense amount of interest in scientists and researchers. Various materials such as nanosensors, nanocarrier drugs, nanoenergetic materials, nanomachines, and nanomaterials showing improved physical and chemical properties, and so on are derived from nanotechnology (Sant, 2012). Increased recognition of the importance of achieving a sustainable environment has led to a growing awareness of the manufacturing and applications of green synthesized nanomaterials. However, relying on non-renewable resources and the generation of hazardous waste persist in many of the processes currently employed. The fusion of nanotechnology with the principles and practices of green chemistry, known as green nanotechnology, offers the potential to build an environmentally sustainable society. Green chemistry emphasizes the design of products and processes that minimize or eliminate the use and generation of hazardous substances. Green nanotechnology practices often involve the utilization of natural sources, nonhazardous solvents, and energy-efficient processes during nanomaterial preparation (Basiuk & Basiuk, 2015). The incorporation of green chemistry principles in current nanomaterial manufacturing processes has facilitated the development of numerous nanoproducts that do not contain toxic substances. Green synthesis of nanomaterials encompasses various approaches for both nano and near-nanomaterials. One objective of green synthesis of nanomaterials is to make them environmentally friendly. Another goal involves leveraging nanotechnology for the development of eco-friendly applications using non-nano materials. The utilization of current materials and processes often relies on non-renewable resources and results in the generation of hazardous waste. To address these challenges, the integration of green chemistry principles with nanotechnology applications has become crucial for the future of nanotechnology. Extensive research has been conducted on the use of natural ingredients and environmentally benign synthetic methods to synthesize nanomaterials (Benelli, 2019). Although only some of these "green nanotechnologies" have transitioned from the lab to pilot-scale applications, they still encounter significant constraints.

DOI: 10.1201/9781003449225-6

6.2 METHODS OF GREEN SYNTHESIS OF NANOMATERIALS

"Nanotechnology is a science in which atoms and molecules make some combinations to form specialized materials". In simpler terms, nanotechnology involves the handling of matter at the nanoscale at the molecular and atomic scale. The concept of nanotechnology was first introduced by Richard Feynman in 1959, marking the beginning of advanced nanotechnology (Hulla et al., 2015). The rapid growth of this emerging field can be seen through the synthesis of nanoscale materials and the exploration of their unique physicochemical properties. Nanotechnology has gained significant momentum in diverse fields including biomedical, chemical, electronics, environmental, food, pharmaceuticals, and optics. Nanomaterials have provided dependable and enduring solutions to various techno-environmental challenges in fields like catalyzing reactions, delivering drugs, treating wastewater, generating hydrogen, and harvesting solar energy.

The global scientific association is actively involved in creating environmentally friendly techniques that synthesize friendly, nontoxic, and highly effective products using green nanotechnology and biotechnology. Green synthesis of nanomaterials in a single-step process ensures environmental compatibility, improved stability, and remarkable characteristics and sizes, disqualifying the requirement for high temperatures, pressures, pH levels, and so on during synthesis.

The nanomaterials synthesis can be broadly classified into three methods: physical, chemical, and biological. These methods can be further categorized as either bottom-up or top-down approaches. Biological, chemical, and physical synthesis methods for nanomaterials typically follow one of these two methodologies (Figure 6.1).

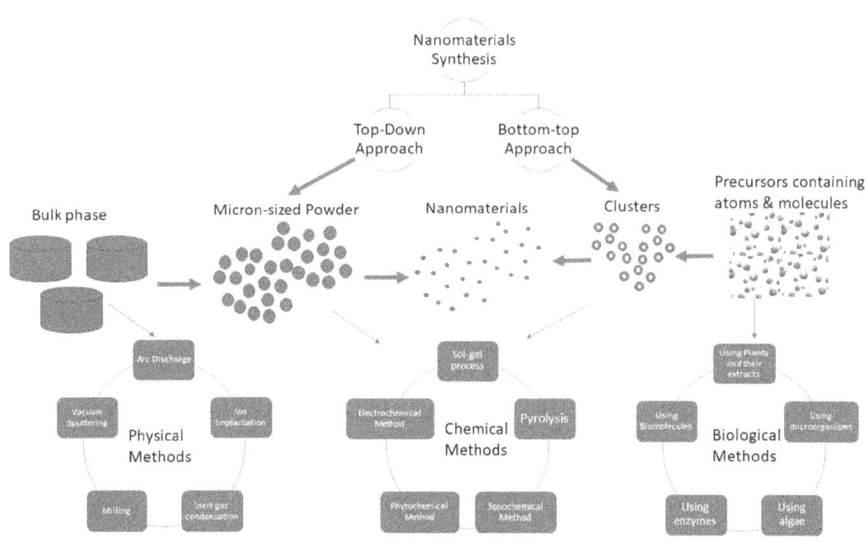

FIGURE 6.1 Methods of nanomaterials synthesis.

6.2.1 Physical Approach

Evaporation-condensation, electrospraying, high-energy ball milling (HEBM), laser ablation, and laser pyrolysis are the most widely employed methods in the physical approach for the synthesis of nanomaterials. Alternative physical approaches include methods such as arc discharge, metal sputtering, atomization, and annealing. Evaporation-condensation methods have been successfully employed to synthesize various nanoparticles, including silver, gold, lead sulfide, and fullerene (Kimoto et al., 1963). These physical approaches offer advantages over chemical methods due to the lack of solvent pollution in preparation of thin films and the even dispersion of synthesized nanoparticles. For instance, silver nanoparticles have been synthesized using a small ceramic heater with localized heating, facilitating rapid cooling and forming of nanomaterials in high concentrations. This physical method finds utility as a nanomaterial synthesizer for extended-duration trials in studies on the toxic effects of inhalation and as a nanomaterial measurement equipment calibration device.

Another physical technique for silver nanoparticle synthesis is laser ablation of metallic bulk materials in a solution. Subdivision of bulk silver metal occurs through laser ablation, where factors such as laser wavelength, pulse duration (femto-, pico-, and nanoseconds), laser fluence (Ullmann et al., 2002), ablation time duration, and the presence or absence of surfactants in the liquid medium affect the efficiency and characteristics of the resulting nanosilver particles. One notable advantage of laser ablation is the absence of chemical reagents in the solution, allowing for the production of pure and uncontaminated metal colloids suitable for various applications.

6.2.2 Chemical Approach

Chemical vapor synthesis, hydrothermal synthesis, microemulsion technique, sol-gel method, and polyol synthesis are the most widely used chemical approaches for the synthesis of nanomaterials. Thermochemical reduction employing materials which possess reducing properties like $NaBH_4$ is the prevailing technique for fabricating silver nanomaterials (Song et al., 2009b). Additional elemental hydrogen, sodium citrate, hydroquinone, and gallic acid are among the reducing agents commonly used in nanoparticle synthesis. Typically, these reactions occur in a solution, yielding a product with characteristics of colloidal systems. As a result, the overall process is commonly referred to as coprecipitation, encompassing concurrent phenomena such as reduction, nucleation, growth, agglomeration, and coarsening.

The above-mentioned reducing agents function by reducing $Ag+$ ions to generate silver metal which subsequently undergoes clustering, forming oligomeric clusters. With time, these clusters further develop into metallic colloidal silver particles. To ensure particle growth stabilization and preservation of surface properties, surfactants containing functional groups like thiols, amines, acids, and alcohols are utilized. These functional groups facilitate interactions with the particle surfaces, preventing sedimentation, agglomeration, or loss of surface properties. While the chemical reduction method enables the synthesis of a substantial quantity of nanoparticles, it is important to note certain drawbacks, including the toxicity of the reagents employed and the generation of environmentally harmful byproducts.

6.2.3 Biological Methods

Bacteria, fungi, plants, and plant parts are employed during nanomaterial synthesis by biological methods. The fundamental requirements for the green synthesis of silver nanoparticles involve a solution containing silver metal ions and a biological agent with reducing properties. In many instances, the reducing agents or other constituents naturally present in cells also serve as stabilizing and capping agents, obviating the need for external addition of such agents. These reducing agents are extensively distributed throughout biological systems. Silver nanoparticles have been successfully synthesized utilizing various organisms from four of the living organisms' classifications, namely monera (prokaryotic organisms lacking a true nucleus), protista (unicellular organisms with a true nucleus), fungi (eukaryotic saprophytes/parasites), and plantae (eukaryotic autotrophs) (Sharma et al., 2009).

6.3 NANOMATERIALS AND ITS TYPES

Nanomaterials lie between the single molecules and their macroscopic counterparts, which have sizes in the range of 1–200 nm. Owing to their small size, nanoparticles possess remarkable physicochemical characteristics, which are characterized by an extensive surface area, elevated energy, and quantum properties. These nanoparticles exhibit a wide range of chemical compositions, including metals (such as silver, gold, copper, zinc), metal oxides, silicates, polymers, organics, and carbon. Furthermore, nanoparticles can adopt various shapes, such as spherical, cylindrical, nanosheets, or wires. This remarkable diversity in shape and chemical composition of nanoparticles is influenced by factors like the medium from which they are formed and the presence of bioactive compounds in that medium.

6.3.1 Types of Nanomaterials

Various types of nanomaterials exist, such as alloy-based, metallic, metal oxide-derived, magnetic, and various other types.

6.3.1.1 Alloy Nanomaterials

Metallic nanomaterial properties can greatly be influenced when combinations of metals are mixed with each other. The combination of different metals in alloy nanoparticles often leads to synergistic effects that enhance specific properties. By adjusting factors such as composition, atomic ordering, and cluster size, it is possible to easily fine-tune the physical and chemical properties of these nanoparticles. This versatility has sparked considerable interest in alloy nanoparticles. Additionally, nanoalloys can exhibit unique properties that are significantly different from their bulk counterparts, opening up numerous applications in fields such as electronics, engineering, and catalysis (Huynh et al., 2020).

6.3.1.2 Metallic Nanomaterials

These nanomaterials are 1–100 nm in size and are called nanometals having nanoscopic dimensions. Metallic nanoparticles possess several significant characteristics, including a high surface-area-to-volume ratio, high surface energy, a typical electronic structure resulting from their jumping from molecular to metallic states, excitation of plasmons, and confinement at the quantum level.

Various categories of metal nanomaterials are prepared, with gold, silver, copper, platinum, and palladium being the primary ones. Silver nanoparticles are particularly favored and widely synthesized through green synthesis methods due to their exceptional antibacterial properties and the ease of reducing from monovalent ions of silver. Silver nanoparticles also exhibit high tendency of plasmon excitation (Song et al., 2009b). They find diverse applications in fields such as medicine, catalysis, textiles, and water treatment. Gold nanomaterials have very considerable importance, albeit not much as compared to silver. While other metal nanomaterials have been reported in the literature, their synthesis remains relatively limited.

6.3.1.3 Metal Oxide Nanomaterials

The synthesis of this nanoparticle type includes connecting metal centers using oxo (M-O-M) or hydroxo (M-OH-M) bridges, resulting in the formation of metal-oxo or metal-hydroxo polymers in solution (Oskam, 2006).

6.3.1.4 Magnetic Nanomaterials

These nanomaterials comprise binary distinct parts: a magnetic component, which can be materials like iron, nickel, or cobalt, along with a chemical component possessing a particular functionality. The inclusion of the magnetic component enables easy manipulation of these nanoparticles using magnetic fields. These nanoparticles exhibit a variety of desirable properties, making them suitable for usages in catalysis, targeted delivery of peculiar tissues, and medical diagnostics (Lu et al., 2007).

6.4 GREEN SOLVENTS

Green solvents, also known as environmentally friendly solvents or sustainable solvents, are solvents that are designed to have reduced environmental impact compared to traditional solvents. They are typically characterized by their low toxicity, low volatility, biodegradability, and non-flammability (Cvjetko Bubalo et al., 2015). Green solvents are important in various industries, including chemical synthesis, pharmaceuticals, coatings, and cleaning products, where solvents are commonly used.

There are several types of green solvents that are commonly employed.

6.4.1 Water

Water is one of the most abundant and environmentally benign solvents. It is nontoxic, nonflammable, and readily available, making it an ideal green solvent for many applications.

6.4.2 Bio-Based Solvents

These solvents are derived from renewable resources, such as plant oils, terpenes, or sugar-based compounds. Examples include ethanol, isopropyl alcohol (derived from bioethanol), limonene, and glycerol. Bio-based solvents can offer reduced toxicity and environmental impact compared to their petroleum-based counterparts.

6.4.3 Supercritical Carbon Dioxide ($scCO_2$)

Carbon dioxide in a supercritical state can act as a solvent under specific temperature and pressure conditions. $scCO_2$ is nontoxic, nonflammable, and readily available. It can be used as a green alternative to organic solvents for various applications.

6.4.4 Ionic Liquids

Ionic liquids consist of salts that exist in a liquid phase or near room temperature. They possess unique features, such as high thermal stability, low volatility, and tunable solvation properties. Ionic liquids are often considered greener alternatives to conventional organic solvents.

6.4.5 Green Organic Solvents

Some organic solvents, such as ethyl lactate, d-limonene, or propylene glycol, exhibit lower toxicity and environmental impact compared to traditional solvents like benzene or toluene. These solvents are derived from renewable resources or have lower volatility, making them greener choices.

The use of green solvents can significantly reduce the environmental footprint associated with various industrial processes, including chemical synthesis, extraction, and cleaning. Green solvents are part of sustainable chemistry practices aimed at minimizing the use of hazardous substances and promoting more environmentally friendly alternatives.

6.5 GREEN SYNTHESIZED NANOMATERIALS

There are various problems such as generation of toxic waste and environmental hazards when nanomaterials are synthesized using physical and chemical methods. The physical method, which involves the use of a tube furnace, has several drawbacks. It requires a significant amount of space, generates excessive heat, leading to an increase in the environmental temperature surrounding the source material, and is time-consuming. On the other hand, the chemical method of nanoparticle generation relies on toxic solvents and chemicals, posing significant harm to the environment. Therefore, there has been a global demand for an alternative approach to nanoparticle generation, leading to the emergence of the concept of green nanotechnology. It offers a simple, cost-effective, and environmentally friendly solution, gaining considerable importance in recent times. Numerous nanoparticles synthesized through green methods have been successfully applied in variety of fields.

Green synthesis primarily involves the use of biological routes such as bacteria, fungi, or plants, along with various biotechnological techniques, for the synthesis of nanomaterials or nanoparticles. The nanoparticles produced through these biological routes are devoid of toxic chemicals and are considered eco-friendly. In the following discussion, we will explore different biological routes in detail for nanoparticle synthesis.

6.5.1 Bacteria-Mediated Nanomaterials

Bacteria can convert a variety of inorganic nanomaterials while synthesizing metallic nanoparticles such as gold and silver. Silver is widely recognized for its antimicrobial properties. However, certain bacteria exhibit resistance to silver and can accumulate it on their cell walls, reaching up to 25% of their dry-weight biomass. This intriguing ability suggests their potential application in the industrial extraction of silver from ore materials. The initial discovery of bacteria synthesizing silver nanoparticles was made through the study of the Pseudomonas stutzeri AG259 strain, which was isolated from a silver mine (Klaus et al., 1999).

The prevailing mechanism for silver biosynthesis involves the presence of the nitrate reductase enzyme, which converts nitrate to nitrite. In the case of bacterial-mediated silver synthesis, the presence of alpha-nicotinamide adenine dinucleotide phosphate reduced form (NADPH)-dependent nitrate reductase eliminates the need for additional downstream processing steps. In a particular study, it was observed that intracellular E. coli DH5α facilitated the synthesis of gold nanoparticles from chloroauric acid. The resulting nanoparticles exhibited predominantly spherical morphology, with some triangular and quasihexagonal shapes observed on the cell surface. This study also revealed the direct electrochemistry of hemoglobin and other proteins (Narayanan & Sakthivel, 2010a). Conversely, the production of magnetite nanoparticles with sizes ranging from 10 to 50 nm through extracellular synthesis was documented under anaerobic conditions. This achievement was accomplished using Geobacter metallireducens GS-15, a nonmagnetotactic bacterium isolated from Potomac River sediments. Likewise, using cobalt acetate as a precursor, crystalline ferromagnetic Co_3O_4 nanoparticles with dimensions of 5–7 nm were extracellularly synthesized. This process involved a metal-tolerant marine bacterium known as *Brevibacterium casei* (Kumar et al., 2009).

6.5.2 Fungi-Mediated Nanomaterials

As fungi show tolerance, high affinity for binding and internalization comparable to bacteria and metal bioaccumulation ability, they are utilized to synthesize metallic nanoparticles. Various types of fungi such as Aspergillus sp., Trichoderma sp., Penicillium sp., Alternaria sp., Trametes sp., Pleurotus sp., Ganoderma sp. have been employed for nanomaterials synthesis. Utilizing fungi for nanoparticle generation offers several advantages compared to other microorganisms, as fungi exhibit faster growth and are easier to handle and cultivate in laboratory processes compared to bacteria. Moreover, the fungal mycelial mesh demonstrates resilience to various conditions encountered in bioreactors or reaction compartments, such as fluid pressure and stirring. In fungi, the mechanism of nanoparticle synthesis differs, as they secrete abundant enzymes that play a role in reducing silver ions, leading to the synthesis of metal nanomaterials. Enzymes released outside the cell, such as naphthoquinones and anthraquinones, are known to facilitate this reduction process. In the case of F. oxysporum, it is believed that nanoparticle formation is attributed to the NADPH-dependent nitrate reductase and a shuttle quinine extracellular process (Prabhu & Poulose, 2012).

The first instance of fungus-mediated nanoparticle synthesis dates back to the early 20th century when Verticillium fungus was used to synthesize silver nanoparticles with an average diameter of 25 ± 12 nm. Verticillium luteoalbum was employed for intracellular gold nanoparticle generation, where the morphology and size of the gold nanoparticles varied based on pH changes in the medium (Gericke & Pinches, 2006). Another example involves the use of an extract from the fungus, such as Volvariella volvacea, for the synthesis of silver and gold combination nanomaterials (Philip, 2009).

6.5.3 Actinomycetes-Mediated Nanomaterials

Most antimicrobial compounds are produced by actinomycetes, which comprise about two-thirds of the total and share various properties with fungi. However, further research is required in the field of nanomaterial synthesis utilizing actinomycetes. Scientists have documented the generation of uniformly sized gold nanomaterials with diameter ~8 nm, utilizing a novel actinomycete called Thermomonospora sp (Sastry et al., 2003). These nanoparticles exhibited remarkable stability for approximately 6 months. Fourier-Transform Infrared Spectroscopic (FTIR) analysis confirmed the presence of proteins responsible for nanoparticle stabilization on the surface of the nanoparticles. Intracellular accumulation of gold nanoparticles, ranging from 5 to 15 nm in size, was observed in an alkalotolerant actinomycete known as Rhodococcus sp. The nanomaterials were observed to be agglomerated on the cell periphery in the form of Ag0 (Ahmad et al., 2003). Similarly, silver nanoparticles have been synthesized by various actinomycetes. These synthesized nanoparticles have been utilized in a wide range of biomedical applications and have demonstrated effective antimicrobial activity against a range of pathogens.

6.5.4 Yeast-Mediated Nanomaterials

Semiconductors are prepared using yeast, which is a unicellular eukaryotic microorganism. Peptide-bound CdS quantum crystallites, which were monodispersed and spherical with a size of 20Å, were synthesized using the yeast Candida glabrata. Additionally, Wurtzite hexagonal CdS crystals were generated by Schizosaccharomyces pombe measuring 1–1.5 nm in size during the mid-log phase. In a similar vein, Torulopsis sp. was utilized to successfully generate intracellular PbS nanocrystallites ranging from 2 to 5 nm in size, which were stored within its vacuoles (Suresh, 2014). These synthesized nanoparticles were then employed in the production of an ideal diode. A silver-tolerant yeast strain known as MKY3 was employed for the extracellular synthesis of silver nanoparticles. When exposed to 1 mM soluble silver during its logarithmic growth phase, this strain produced silver nanoparticles ranging from 2 to 5 nm in size. The particles were subsequently the sample was separated by exploiting differential thawing. The silver nanoparticles were surely formed through various techniques including optical absorption, transmission electron microscopy, X-ray diffraction, and X-ray photoelectron spectroscopy.

6.5.5 Algae-Mediated Nanomaterials

Nanomaterials are synthesized using algae in many of the studies done. Protein-mediated synthesis of gold nanoparticles using the blue-green alga Spirulina platensis has been reported, resulting in the production of uniformly sized nanomaterials average sized ~5 nm. These nanoparticles were employed to conduct assays to evaluate the antibacterial activity against Bacillus subtilis and Staphylococcus aureus. Sargassum wightii has demonstrated the ability to produce stabilized and well-distributed gold nanomaterials in the size range of 8–12 nm. This was confirmed through UV-Vis spectroscopy, TEM, and XRD techniques. In the same way, Sargassum wightii was also used as the source of silver nanoparticles and was tested for its antibacterial activity (Shanmugam et al., 2014). Diatoms such as Navicula atomus and Diadesmis gallica have been utilized to synthesize gold nanoparticles, EPS-gold, and silica-gold bionanocomposites.

6.5.6 Biomolecular Templates Nanomaterials

Various nanomaterials are synthesized using biomolecules such as cell membranes, nucleic acids, and viruses. DNA, due to its high attraction for transition metals, serves as a perfect biomolecular template. Studies have demonstrated gold nanoparticle synthesis by creating and cross-linking a DNA hydrogel, followed by the incorporation of transition metal ions like gold into DNA macromolecules. In this process, bioreduction of Au3+ ions to Au0 atoms occurs, resulting in the formation of metal clusters on the DNA chain that develop into Au nanoparticles. These nanoparticles have been utilized for catalytic activities. DNA has also been employed as a template for the synthesis of highly stable, wire-like clusters of silver nanoparticles, which were subsequently used as ultrasensitive substrates for surface-enhanced Raman scattering (SERS). Additionally, a method using DNA as a template and an electroless photolytic approach has been devised for the synthesis of gold nanostructures. These nanoclusters range in size from 10 to 40 nm, while the nanostructures have diameters of 40–70 nm and exhibit resistivity comparable to pure metal. They are continuous and electrically conductive, displaying low electrical resistance and strong adherence to electrodes. DNA is both a capping and a reducing agent, used in conjunction with UV irradiation, to synthesize electrically conducting nanowires of Au, Pd, and CdS. These nanowires have potential applications as building blocks for various devices of nanosize, nanosensors, and electronic devices. Biological membranes, with their ultrafine pores, have been employed as templates for synthesizing and engineering nanoparticles. For example, gold nanoparticles have been manufactured utilizing a rubber membrane derived from Hevea brasiliensis, which acts as a preservative while reducing gold ion solution at ~80°C (Santos et al., 2019). Similarly, viruses have been utilized as templates to synthesize nanoparticles with uniform size and morphology, utilizing the hollow spaces within their structures. By using tobacco mosaic virus (TMV) templates, iron oxides, CdS and PbS cocrystallizations, and sol-gel condensations of SiO_2 have been synthesized. Additionally, the M13 bacteriophage has been used for the synthesis of quantum dots of ZnS and CdS.

6.5.7 Plant-Mediated Nanomaterials

The utilization of plants in the production and assembling of silver nanoparticles has gained significant interest in recent years. This approach offers a rapid, eco-friendly, biocompatible, and cost-effective protocol, providing a single-step technique for the biosynthetic process. The combination of biomolecules found in plant extracts, including proteins, amino acids, polysaccharides, terpenes, alkaloids, phenolics, saponins, and vitamins, enables the reduction and stabilization of silver ions, making it the easiest and most cost-effective method for producing silver nanoparticles (Figure 6.2).

The use of plant extracts for silver nanoparticle synthesis offers several advantages. Plant extracts are readily available, safe, and generally nontoxic. They contain a wide range of small molecules produced as intermediates or end products of metabolism that boost the reduction of silver ions and facilitate quicker synthesis compared to microbes. The photochemicals present in plants play a crucial role in the reduction process. Steroids, saponins, carbohydrates, and flavonoids found in plant extracts act as reducing agents, while phytoconstituents serve as capping agents, providing stability to the silver nanoparticles. Table 6.1 provides a summary of nanomaterials synthesized using green synthesis methods.

The kinetics of plant-mediated nanomaterial synthesis are significantly higher than other biosynthetic methods, including chemical synthesis. Silver nanomaterials

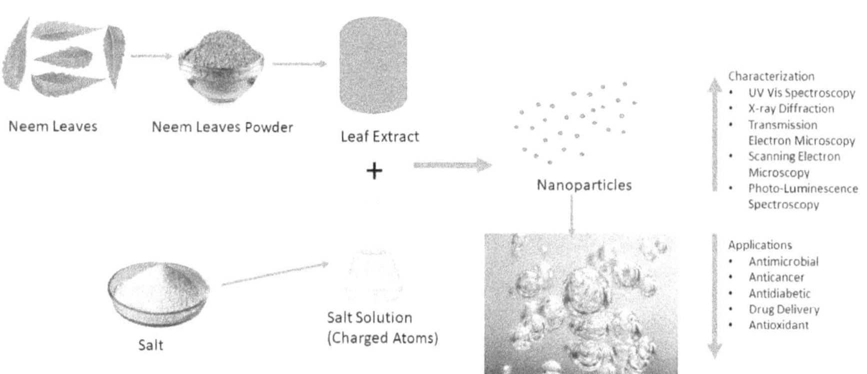

FIGURE 6.2 Green synthesis of nanomaterials.

TABLE 6.1
Nanomaterials Synthesized Using Various Bioresources

Sr. No.	Nanomaterial	Size	Bioresource	Applications	References
1.	Silver	35–55 nm	Catharanthus roseus	Antibacterial and antimicrobial	(Ponarulselvam et al., 2012)
2.	Silver	27 nm	Phlomis leaf extract	Antibacterial and antimicrobial	(Allafchian et al., 2016)
3.	Silver	~ 98 nm	Saraca indica	Antibacterial and antimicrobial	(Perugu et al., 2016)
4.	Zinc Oxide	13–61 nm	Moringa oleifera	UV protection, Antibacterial and Antifungal	(Matinise et al., 2017)
5.	Zinc Oxide	12–53 nm	Limonia acidissima	UV protection, Antibacterial and Antifungal	(Patil & Taranath, 2016)
6.	Gold	(5–300 nm)	Magnolia kobus and Diopyros kaki	Drug delivery, Biosensors, Catalysis	(Song et al., 2009a)
7.	Gold	4.6 to 55.1 nm	Coleus amboinicus Lour	Drug delivery, Biosensors, Catalysis	(Narayanan & Sakthivel, 2010b)
8.	Iron	~50 nm	Sorghum (plant) source bran	Drug delivery, Catalysis	(Njagi et al., 2011)
9.	Iron	15–50 nm	Green tea and eucalyptus leaves extracts	Drug delivery, Catalysis	(Wang et al., 2014)
10.	Copper	97 nm	Murraya koenigii leaf	Conductive inks, Catalysts, Anti-fouling coat	(Mohanraj et al., 2016)
11.	Copper	90 nm	Ginkgo biloba L. leaf	Conductive inks, Catalysts, Anti-fouling coat	(Nasrollahzadeh & Mohammad Sajadi, 2015)

have been synthesized using medicinal plants including Oryza sativa, Helianthus annus, Saccharum officinarum, Sorghum bicolor, Aloe vera, Zea mays, Basella alba, and Capsicum annuum, which have found applications in pharmaceutical and biological fields. For instance, the aqueous extract of Alternanthera dentate rapidly synthesized antibacterial silver nanoparticles ranging from 50 to 100 nm in size within 10 minutes (Kumar et al., 2014). These nanoparticles exhibited antibacterial activity against various pathogens. Gold nanoparticles with morphologies such as spherical, triangular, hexagonal, and rod-shaped were synthesized using Cymbopogon citratus leaf extract and were found to enhance the predation efficiency of copepod Mesocyclops aspericornis against malaria and dengue mosquitoes. Copper oxide nanoparticles were synthesized extracellularly using the medicinal plant Euphorbia nivulia through a colloid thermal synthesis process, and their stability was attributed to the presence of terpenoids and peptides in the plant's latex. Antibacterial activity of the particles was also observed. Indium oxide (In_2O_3) nanoparticles were synthesized using indium acetylacetonate solution extracted from Aloe vera, and their structural, morphological, and optical properties were characterized. Using Sedum alfredii, which is a zinc hyperaccumulator plant, we synthesized zinc oxide nanoparticles with a hexagonal wurtzite structure. These particles had an average size of 53.7 nm (Qu et al., 2011). Additionally, iron oxide nanoparticles were produced using alfalfa at various pH values, resulting in aggregates ranging from 1 to 10 nm in size. Optimal pH conditions allowed for the control of nanoparticle size in the range of 1 to 4 nm. Some of the green synthesized metal oxide nanomaterials are depicted in Figure 6.3.

FIGURE 6.3 Green synthesized metal oxide nanomaterials.

Source: Polshettiwar and Varma (2010).

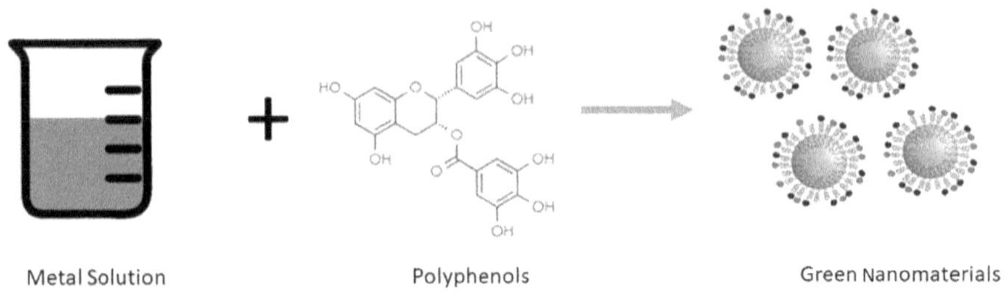

FIGURE 6.4 Mechanism of green synthesis from bioresources.

6.6 MECHANISM INVOLVED IN GREEN NANOMATERIALS SYNTHESIS

Methods of synthesis of nanomaterials decide the mechanisms as to what is taking place at molecular and atomic levels. Nanoparticle synthesis encompasses the generation of metallic iron nanoparticles that serve as both reducing and capping agents, enabling the production of iron particles at the nanoscale for various applications. Phytochemical screening, including tests for alkaloids, flavonoids, phenols, tannins, quinones, saponins, and other compounds, is conducted to identify the presence of polyphenols. The different phytochemical compounds present in plant extracts possess distinct reducing properties, which play a role in controlling the nanoparticle synthesis process by modulating the precursor. Phenol derivatives and flavonoids, in particular, are crucial for their biological effects, such as antioxidant and free-radical properties.

The synthesis of nanomaterials from plant extracts involves three phases such as activation phase, growth phase, and process termination phase. Detailed mechanisms and processes involved in the synthesis are depicted in Figure 6.4. In the activation phase, metal ion reduction occurs, forming newer complexes through self-nucleation of the reduced metal atoms. This phase involves growing the newly formed structures and further reducing metal ions, resulting in an increase in thermodynamic stability. As the nanoparticles reach the third phase, their shape will determine their stability.

6.7 FACTORS AFFECTING GREEN SYNTHESIS OF NANOMATERIALS

Lot of research has taken place as to what is the mechanism behind nanomaterials synthesis, which is very complicated. However, the majority of available literature suggests that biomolecules, either individually or in combination, play a crucial role as reducing and capping agents in nanoparticle synthesis. Several factors, including temperature, particle size, pH, and extract concentration, may influence the synthesis process.

The variation in nanoparticle size can be attributed to two possible reasons: (1) the concentration of polyphenols, as they have a significant impact on the reducing and capping performance; and (2) pH alterations, where acidic or alkaline pH ranges can lead to agglomeration through excessive nucleation (at low pH) or nanoparticle instability (at high pH).

Experimental investigation of parameters, such as the initial reaction mixture's pH, enables easy control over the size and shape of nanomaterials. pH variation strongly influences the nucleation process. Overall, acidic conditions promote nanoparticle formation but with poor stabilization. On the other hand, alkaline conditions facilitate rapid and stable nanoparticle production. Under acidic conditions, nanoparticle growth tends to form clusters that agglomerate, while basic conditions lead to the appearance of numerous pearl-like nanoparticles along with some larger-diameter nanoparticles. The resulting nanoparticles are intended to exhibit stability under a wide range of environmental conditions and find application in various fields.

Higher temperatures (above room temperature) promote faster growth dynamics but also increase the occurrence of defects, thereby affecting the crystal quality. Nucleation time is another critical factor for controlling nanoparticle size and size distribution, as shorter nucleation times result in better size control.

Optimization of parameters such as the solvent ratio, reaction time, temperature, pH, and the mixing ratio of the metal-ion solution to plant juice is essential. However, the interaction between polyphenols and nanoparticles remains unclear due to the involvement of different biomolecules responsible for the reducing and capping agent functions.

Surface charge is another important parameter for nanoparticle characterization. Electrostatic interactions between bioactive compounds from bacteria, plants, algae, and fungi depend on the nature and strength of the particle's surface charge.

Various factors, including pH, temperature, and reaction time, play a critical role in controlling the synthesis and stabilization of nanoparticles produced by biological entities.

Some of the factors that influence the green synthesis of nanomaterials are discussed in the following sections.

6.7.1 pH of the Reaction Medium

The pH of the reaction medium is a critical factor in nanoparticle generation, impacting their size and shape. Variation in hydrogen ion concentrations results in differences in nanoparticle characteristics. Lower acidic pH values tend to produce larger nanoparticles compared to higher pH values. For instance, Avena sativa facilitated the formation of large rod-shaped gold nanoparticles (25–85 nm) at pH 2, while relatively smaller nanoparticles (5–20 nm) were synthesized at pH 3 and 4. Similarly, Cinnamon zeylanicum bark extract led to the synthesis of more spherical silver nanoparticles at higher pH values (pH 5 and above) (Sathishkumar et al., 2009).

6.7.2 Reaction Temperature

The temperature significantly influences the synthesis of metallic nanoparticles, particularly their shapes and sizes. For instance, during the synthesis of gold nanoparticles using Cymbopogon flexosus leaf extract, lower reaction temperatures favored the formation of nanotriangles (Pathania et al., 2022). However, at higher reaction temperatures, more spherical nanoparticles were synthesized alongside the nanotriangles.

6.7.3 Production Technique

Nanoparticles can be synthesized using various methods, including physical, chemical, and biological approaches, each with its own advantages and disadvantages. Among these methods, biological synthesis is highly regarded due to its eco-friendly and nontoxic nature, making it superior to other synthetic routes.

6.7.4 Pressure

The role of pressure in metallic nanoparticle synthesis is significant, as it influences the size and shape of the resulting nanoparticles. Research indicates that under ambient pressure conditions, the reduction of metal ions using phytochemical agents occurs at an accelerated rate compared to normal conditions.

6.7.5 Time

The duration of nanoparticle reaction media incubation significantly impacts the quality and morphology of the nanoparticles. Time variations, along with factors such as light exposure, synthesis method, and storage conditions, also influence the properties of the synthesized nanoparticles. Prolonged incubation can lead to aggregation or shrinkage, potentially diminishing the nanoparticles' effectiveness.

6.8 APPLICATIONS OF GREEN SYNTHESIZED NANOMATERIALS

Nanomaterials have wide-ranging applications in fields such as medicine, chemistry, environment, energy, agriculture, information technology, communication, heavy industry, and consumer goods. Figure 6.5 illustrates some of the nanoparticles synthesized using different metal solutions and their potential applications. Specifically, nanomaterials hold significant promise for surface and groundwater treatment, wastewater management, and remediation of contaminated sites.

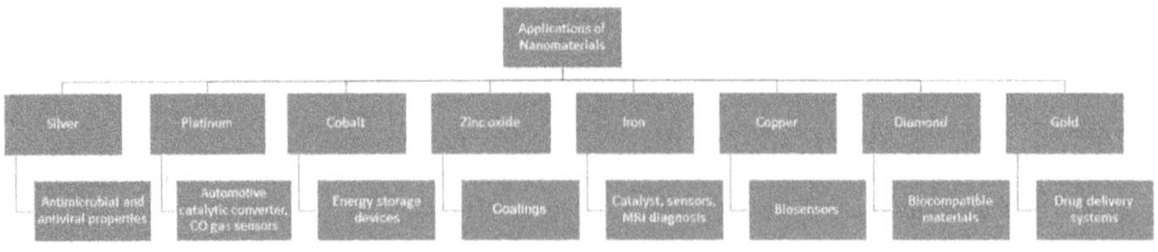

FIGURE 6.5 Applications of various nanomaterials

6.9 LIMITATIONS OF GREEN SYNTHESIZED NANOMATERIALS

The field of green synthesis of nanoparticles is still in its early stages and requires further optimization to ensure desired particle characteristics and effective delivery into target organisms. Various synthesis methods have been explored in the literature to obtain green nanomaterials with specific sizes, shapes, and functionalities. However, there are several limitations associated with green synthesis methods, including the following:

1. Limited studies have been conducted to assess the sustained stability of synthesized green nanomaterials in diverse physical environments such as extreme temperatures, pressures, and pH conditions.
2. Although sustained release of green nanomaterials has been achieved, the development of materials that exhibit responsive release based on environmental changes is still underway. In particular, the pesticide industry is actively researching environmentally responsive, targeted, and controlled-release formulations.
3. The bioaccumulation and toxicity aspects of green nanomaterials in the environment have been insufficiently explored, and further studies are needed to better understand their impact.
4. Green synthesis methods have not gained widespread popularity in industries compared to chemically synthesized nanoparticles. Increased investment in research through federal grants is expected to promote and encourage further advancements in this direction.

Overall, addressing these limitations and conducting more extensive research will contribute to the development and utilization of green nanomaterials with enhanced stability, responsiveness, reduced toxicity, and wider industrial adoption. In addition, few challenges have been listed in Table 6.2.

6.10 FUTURE PERSPECTIVES AND RESEARCH DIRECTIONS

The research on green synthesis of nanomaterials depends on the identification of newer biological materials for keeping pace with the market. Further efforts are necessary to validate the relatively nontoxic nature of green nanomaterials. This includes the establishment of government regulatory bodies and the conduct of clinical trials. Future optimization of green-synthesized nanomaterials should consider the following:

1. Screening various organisms from different ecosystems as potential sources of biomolecules for green nanomaterials synthesis. For example, exploring thermophilic fungi as mediators for green nanomaterials synthesis, which may exhibit improved thermal stability.

TABLE 6.2
Constraints While Scaling-Up the Concepts of Green Synthesis into Reality

Sr. No.	Key Challenges	Reasons
1.	Undefined standard operating procedures	Researchers in the early stages of exploring green nanoscience lack specific design guidelines.
2.	Create and innovate novel technology	Research in basic science, engineering, and coordination between industrial and research communities is necessary for the commercial production of green nanomaterials.
3.	Lack of expertise	The limited availability of experienced scientists and engineers skilled in the development of green nanotechnology.
4.	Biologically not compatible	Protocols for toxicology and analysis should be developed and regularly updated to align with the latest advancements in the field.
5.	Rigorous safety precautions	The regulatory environment for green nanotechnology remains uncertain, and these technologies are often subject to higher barriers to entry than traditional synthetic chemicals.
6.	Absence of marketing management opportunities	The demand in the final market is uncertain, particularly due to the limited availability of commercially viable products that can be compared to conventional materials in terms of performance.

2. Identifying the molecular components involved in the chemical reduction, nucleation, growth, and stabilization of green nanomaterials, as well as their active role in antimicrobial action.
3. Assessing the toxicity of green nanomaterials compared to chemically synthesized nanomaterials using multiple plant and animal species as models for new applications.
4. Comparing the antimicrobial efficacy of green nanomaterials with nanomaterials synthesized through conventional chemical and physical methods.
5. Investigating the long-term effects of green nanomaterials on multidrug-resistant species using the lowest possible concentrations.
6. Exploring synthetic approaches, such as CRISPR-Cas9-mediated modulation of the genetic architecture of microbes or plants, to enhance the production of size and shape-tuned nanomaterials. Identifying industrially feasible strains through genome-scale manipulation and selection is also important.
7. Conducting analyses on the pharmacokinetics and pharmacodynamics of green nanomaterials against different surfaces and materials.

6.11 CONCLUSIONS

Nanomaterials have very unique physical and chemical properties and hence are very important today. Green synthesis of nanoparticles offers a simple, cost-effective, and environmentally friendly approach, enabling their production in a short time with minimal effort and without causing harm to the environment. The current focus on environmental protection emphasizes the importance of green practices, such as recycling and cost-effective reuse, to enhance quality of life and health. Extensive research is dedicated to exploring the synthesis, characteristics, and behavior of nanomaterials, as well as their applications in diverse fields including medicine, pharmacology, water treatment, agriculture, and wastewater pollutant degradation.

Considering the vast potential of nanoparticle applications in different fields, the choice of synthesis technique should be based on locally available resources and their techno-economic feasibility. Furthermore, a comprehensive understanding of biochemical pathways will contribute to the sustainable development of green synthesis methods.

REFERENCES

Ahmad, A., Senapati, S., Khan, M. I., Kumar, R., Ramani, R., Srinivas, V., & Sastry, M. (2003). Intracellular synthesis of gold nanoparticles by a novel alkalotolerant actinomycete, Rhodococcus species. *Nanotechnology*, *14*(7), 824. https://doi.org/10.1088/0957-4484/14/7/323

Allafchian, A. R., Mirahmadi-Zare, S. Z., Jalali, S. A. H., Hashemi, S. S., & Vahabi, M. R. (2016). Green synthesis of silver nanoparticles using phlomis leaf extract and investigation of their antibacterial activity. *Journal of Nanostructure in Chemistry*, *6*(2), 129–135. https://doi.org/10.1007/s40097-016-0187-0

Basiuk, V. A., & Basiuk, E. V. (2015). Green processes for nanotechnology: From inorganic to bioinspired nanomaterials. *Green Processes for Nanotechnology: from Inorganic to Bioinspired Nanomaterials*, 1–446. https://doi.org/10.1007/978-3-319-15461-9

Benelli, G. (2019). Green synthesis of nanomaterials. *Nanomaterials*, *9*(9), 1275. https://doi.org/10.3390/nano9091275

Cvjetko Bubalo, M., Vidović, S., Radojčić Redovniković, I., & Jokić, S. (2015). Green solvents for green technologies. *Journal of Chemical Technology & Biotechnology*, *90*(9), 1631–1639. https://doi.org/10.1002/jctb.4668

Gericke, M., & Pinches, A. (2006). Microbial production of gold nanoparticles. *Gold Bulletin*, *39*(1), 22–28. https://doi.org/10.1007/BF03215529

Hulla, J., Sahu, S., & Hayes A.. (2015). *Human & Experimental Toxicology*, *34*(12), 1318–1321. doi:10.1177/0960327115603588

Huynh, K. H., Pham, X. H., Kim, J., Lee, S. H., Chang, H., Rho, W. Y., & Jun, B. H. (2020). Synthesis, properties, and biological applications of metallic alloy nanoparticles. *International Journal of Molecular Sciences*, *21*(14), 5174. https://doi.org/10.3390/IJMS21145174

Kimoto, K., Kamiya, Y., Nonoyama, M., & Uyeda, R. (1963). An electron microscope study on fine metal particles prepared by evaporation in argon gas at low pressure. *Japanese Journal of Applied Physics*, *2*(11), 702–713. https://doi.org/10.1143/JJAP.2.702

Klaus, T., Joerger, R., Olsson, E., & Granqvist, C.-G. (1999). Silver-based crystalline nanoparticles, microbially fabricated. *Proceedings of the National Academy of Sciences*, *96*(24), 13611–13614. https://doi.org/10.1073/pnas.96.24.13611

Kumar, D. A., Palanichamy, V., & Roopan, S. M.. (2014). Green synthesis of silver nanoparticles using Alternanthera dentata leaf extract at room temperature and their antimicrobial activity. *Spectrochimica Acta Part A*, *127*, 168–171. www.sciencedirect.com/science/article/pii/S1386142514002455

Kumar, U., Vivekanand, K., & Poddar, P. (2009). Real-time nanomechanical and topographical mapping on live bacterial cells—*Brevibacterium casei* under stress due to their exposure to Co^{2+} ions during microbial synthesis of Co_3O_4 nanoparticles. *The Journal of Physical Chemistry B*, *113*(22), 7927–7933. https://doi.org/10.1021/jp902698n

Lu, A.-H., Salabas, E. L., & Schüth, F. (2007). Magnetic nanoparticles: synthesis, protection, functionalization, and application. *Angewandte Chemie International Edition*, *46*(8), 1222–1244. https://doi.org/10.1002/anie.200602866

Matinise, N., Fuku, X. G., Kaviyarasu, K., Mayedwa, N., & Maaza, M. (2017). ZnO nanoparticles via Moringa oleifera green synthesis: Physical properties and mechanism of formation. *Applied Surface Science*, *406*, 339–347. https://doi.org/10.1016/J.APSUSC.2017.01.219

Mohanraj, S., Anbalagan, K., Rajaguru, P., & Pugalenthi, V. (2016). Effects of phytogenic copper nanoparticles on fermentative hydrogen production by Enterobacter cloacae and Clostridium acetobutylicum. *International Journal of Hydrogen Energy*, *41*(25), 10639–10645. https://doi.org/10.1016/j.ijhydene.2016.04.197

Narayanan, K. B., & Sakthivel, N. (2010a). Biological synthesis of metal nanoparticles by microbes. *Advances in Colloid and Interface Science*, *156*(1–2), 1–13. https://doi.org/10.1016/J.CIS.2010.02.001

Narayanan, K. B., & Sakthivel, N. (2010b). Phytosynthesis of gold nanoparticles using leaf extract of Coleus amboinicus Lour. *Materials Characterization*, *61*(11), 1232–1238. https://doi.org/10.1016/j.matchar.2010.08.003

Nasrollahzadeh, M., & Mohammad Sajadi, S. (2015). Green synthesis of copper nanoparticles using Ginkgo biloba L. leaf extract and their catalytic activity for the Huisgen [3 + 2] cycloaddition of azides and alkynes at room temperature. *Journal of Colloid and Interface Science*, *457*, 141–147. https://doi.org/10.1016/j.jcis.2015.07.004

Njagi, E. C., Huang, H., Stafford, L., Genuino, H., Galindo, H. M., Collins, J. B., Hoag, G. E., & Suib, S. L. (2011). Biosynthesis of iron and silver nanoparticles at room temperature using Aqueous Sorghum bran extracts. *Langmuir*, *27*(1), 264–271. https://doi.org/10.1021/la103190n

Oskam, G. (2006). Metal oxide nanoparticles: synthesis, characterization and application. *Journal of Sol-Gel Science and Technology*, *37*(3), 161–164. https://doi.org/10.1007/s10971-005-6621-2

Pathania, D., Sharma, M., Thakur, P., Chaudhary, V., Kaushik, A., Furukawa, H., & Khosla, A. (2022). Exploring phytochemical composition, photocatalytic, antibacterial, and antifungal efficacies of Au nanomaterials supported by Cymbopogon flexuosus essential oil. *Scientific Reports*, *12*(1), 14249. https://doi.org/10.1038/s41598-022-15899-9

Patil, B. N., & Taranath, T. C. (2016). Limonia acidissima L. leaf mediated synthesis of zinc oxide nanoparticles: A potent tool against Mycobacterium tuberculosis. *International Journal of Mycobacteriology*, *5*(2), 197–204. https://doi.org/10.1016/j.ijmyco.2016.03.004

Perugu, S., Nagati, V., & Bhanoori, M. (2016). Green synthesis of silver nanoparticles using leaf extract of medicinally potent plant Saraca indica: a novel study. *Applied Nanoscience*, *6*(5), 747–753. https://doi.org/10.1007/s13204-015-0486-7

Philip, D. (2009). Biosynthesis of Au, Ag and Au–Ag nanoparticles using edible mushroom extract. *Spectrochimica Acta Part A: Molecular and Biomolecular Spectroscopy*, *73*(2), 374–381. www.sciencedirect.com/science/article/pii/S1386142509001103

Polshettiwar, V., & Varma, R. S. (2010). Green chemistry by nano-catalysis. *Green Chemistry*, *12*(5), 743. https://doi.org/10.1039/b921171c

Ponarulselvam, S., Panneerselvam, C., Murugan, K., Aarthi, N., Kalimuthu, K., & Thangamani, S. (2012). Synthesis of silver nanoparticles using leaves of Catharanthus roseus Linn. G. Don and their antiplasmodial activities. *Asian Pacific Journal of Tropical Biomedicine*, *2*(7), 574. https://doi.org/10.1016/S2221-1691(12)60100-2

Prabhu, S., & Poulose, E. K. (2012). Silver nanoparticles: mechanism of antimicrobial action, synthesis, medical applications, and toxicity effects. *International Nano Letters*, *2*(1), 32. https://doi.org/10.1186/2228-5326-2-32

Qu, J., Luo, C., & Hou, J. (2011). Synthesis of ZnO nanoparticles from Zn-hyperaccumulator (Sedum alfredii Hance) plants. *Micro & Nano Letters*, *6*(3), 174. https://doi.org/10.1049/mnl.2011.0004

Sant, S. B. (2012). Nanoparticles: From theory to applications. *Materials and Manufacturing Processes*, *27*(12), 1462–1463. https://doi.org/10.1080/10426914.2012.663137

Santos, N. M., Gomes, A. S., Cavalcante, D. G. S. M., Santos, L. F., Teixeira, S. R., Cabrera, F. C., & Job, A. E. (2019). Green synthesis of colloidal gold nanoparticles using latex from *Hevea brasiliensis* and evaluation of their in vitro cytotoxicity and genotoxicity. *IET Nanobiotechnology*, *13*(3), 307–315. https://doi.org/10.1049/iet-nbt.2018.5225

Sastry, M., Ahmad, A., Khan, M. I., & Kumar, R.. (2003). Biosynthesis of metal nanoparticles using fungi and actinomycete. *JSTOR*, *85*(2). www.jstor.org/stable/24108579

Sathishkumar, M., Sneha, K., Won, S. W., Cho, C.-W., Kim, S., & Yun, Y.-S. (2009). Cinnamon zeylanicum bark extract and powder mediated green synthesis of nano-crystalline silver particles and its bactericidal activity. *Colloids and Surfaces B: Biointerfaces*, *73*(2), 332–338. https://doi.org/10.1016/j.colsurfb.2009.06.005

Shanmugam, N., Rajkamal, P., Cholan, S., Kannadasan, N., Sathishkumar, K., Viruthagiri, G., & Sundaramanickam, A. (2014). Biosynthesis of silver nanoparticles from the marine seaweed Sargassum wightii and their antibacterial activity against some human pathogens. *Applied Nanoscience*, *4*(7), 881–888. https://doi.org/10.1007/s13204-013-0271-4

Sharma, V. K., Yngard, R. A., & Lin, Y. (2009). Silver nanoparticles: Green synthesis and their antimicrobial activities. *Advances in Colloid and Interface Science*, *145*(1–2), 83–96. https://doi.org/10.1016/j.cis.2008.09.002

Song, J. Y., Jang, H.-K., & Kim, B. S. (2009a). Biological synthesis of gold nanoparticles using Magnolia kobus and Diopyros kaki leaf extracts. *Process Biochemistry*, *44*(10), 1133–1138. https://doi.org/10.1016/j.procbio.2009.06.005

Song, K. C., Lee, S. M., Park, T. S., & Lee, B. S. (2009b). Preparation of colloidal silver nanoparticles by chemical reduction method. *Korean Journal of Chemical Engineering*, *26*(1), 153–155. https://doi.org/10.1007/S11814-009-0024-Y

Suresh, A. K. (2014). Extracellular bio-production and characterization of small monodispersed CdSe quantum dot nanocrystallites. *Spectrochimica Acta Part A: Molecular and Biomolecular Spectroscopy*, *130*, 344–349. https://doi.org/10.1016/j.saa.2014.04.021

Ullmann, M., Friedlander, S. K., & Schmidt-Ott, A. (2002). Nanoparticle formation by laser ablation. *Journal of Nanoparticle Research*, *4*(6), 499–509. https://doi.org/10.1023/A:1022840924336

Wang, T., Lin, J., Chen, Z., Megharaj, M., & Naidu, R. (2014). Green synthesized iron nanoparticles by green tea and eucalyptus leaves extracts used for removal of nitrate in aqueous solution. *Journal of Cleaner Production*, *83*, 413–419. https://doi.org/10.1016/j.jclepro.2014.07.006

7 Exploring the Components of Green Building Envelopes for Sustainable Civil Infrastructure
A Critical Review

Ritu Bala Garg and Gurpreet Singh

7.1 INTRODUCTION

The building and construction industry is responsible for up to 38% of annual global greenhouse gas (GHG) emissions. Construction industry is putting positive and negative impact on the environment by its operation. Construction and demolition waste, e-waste (Aggarwal, Rana, and Akha, 2022; Akhai, 2023a), the consumption of huge number of resources for construction, creating air and noise pollution while running the diesel-powered equipment for the implementation (Goyal, 2023). Lot of energy is consumed resulting in carbon emissions, thus causing global warming (Lu, Cui, and Li, 2016; Kumar and Akhai, 2022). Energy use and carbon emissions occur in all stages of the life cycle of a building, which is known as the building's carbon footprint (Mneimneh, Ghazzawi, and Ramakrishna, 2023). These carbon emissions may be divided into general groups such as Embodied Energy, Operational Energy, Life-Cycle Energy, Renewable Energy, and Peak Energy (Asdrubali et al., 2023; Song, Zhang, and Mithraratne, 2023; Li and Tingley, 2023; Christopher et al., 2023). The embodied energy of a product or material refers to the total amount of energy consumed during its manufacturing, transportation, and implementation. Building materials are a major contributor to embodied energy consumption. On the other hand, operational energy is associated with the energy used by a building during its occupancy stage, providing thermal and visual comfort to occupants. Renewable energy is a naturally replenished form of energy that can be a part of a building's energy profile. Throughout its lifetime, a building goes through various stages, and the sum of all energy associated

DOI: 10.1201/9781003449225-7

with these stages, including embodied, operational, and renewable energy, results in the building's life-cycle energy (Koura et al., 2020; Füchsl, Rheude, and Röder, 2022; Tavares et al., 2021). These all vary considerably depending on the type and functionality of the building, climate, location, orientation, building massing, and so on (Gao et al., 2023; Pan et al., 2023). For human comfort conditions, air conditioners are used in buildings (Akhai, Mala, and Jerin, 2021; Tanwar and Akhai, 2017; Akhai, Mala, and Jerin, 2020). These devices consume large amount of electricity (Mathias, Juenger, and Horton, 2023; Akhai, Thareja, and Singh, 2017; Alam et al., 2023; Gorantla, Shaik, and Setty, 2017) and majorly contribute to energy consumption and global warming (Akhai, Bansal, and Singh, 2020; Kaushal, Singh, and Gupta, 2022; Kaushal, Gupta, and Bhowmick, 2019; Kaushal et al., 2019). Due to the high consumption of energy as well as carbon emissions, Nearly Zero Energy Buildings (NZEB) concepts are examined for new residential and office buildings using state-of-the-art technologies already available on the market. Figure 7.1 presents an overview of the fundamental concept of a zero energy building. Figure 7.2 provides a depiction of the essential components that constitute a zero energy building.

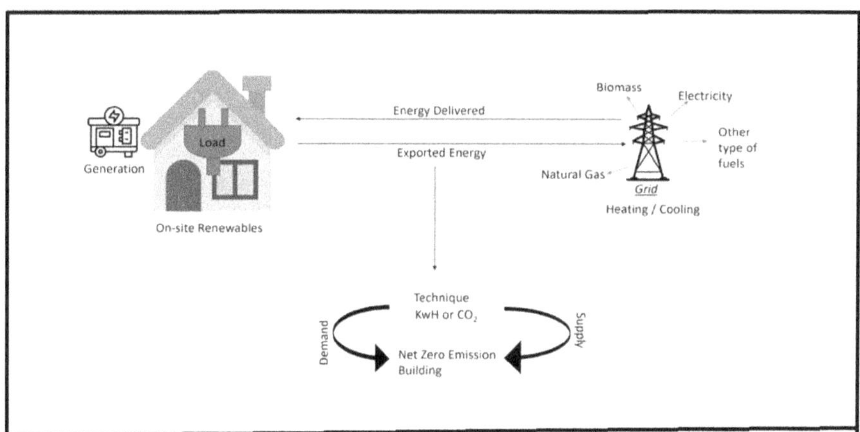

FIGURE 7.1 Basic concept of zero energy building.

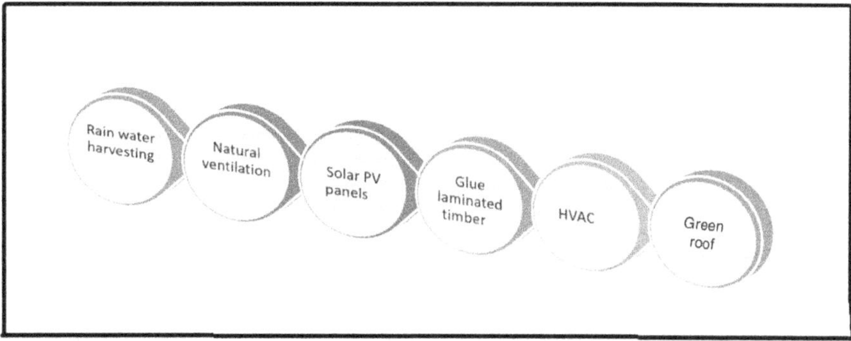

FIGURE 7.2 Essential components of zero energy building.

To achieve the targets of NZEB, the building's outer leaf, designed with high-performance materials, aims to reduce energy consumption. Essential components of a building envelope (Goyal and Dhanoa, 2022; Goyal, 2023) often include the following:

- Foundation
- Wall assemblies
- Roofing systems
- Doors and doorways
- Other components, such as chimneys, vents, and windows

7.2 EXISTING WORK ON ZERO ENERGY BUILDINGS

The components of zero energy buildings, such as fenestration, walls, and roof, are essential to achieving zero energy consumption. Figure 7.3 illustrates the various elements of a zero energy building. Considering the effect of roofing material on embodied energy mentioned that by implementing PV panels, the configuration of the proposed roof system can significantly improve energy efficiency and economic, social, and environmental benefits (Pragati et al., 2023; Mihalakakou et al., 2023). Not only the energy efficiency even carbon and embodied energy should also be considered as sheet metal roofs emit 73% of GHG and have 71% of embodied energy demand (Sodagar et al., 2011; Khan and Baqi, 2021). The type of roof form, shape, and thickness plays a vital role; for example, non-flat structural roof forms reduce the embodied energy by up to 40% (Chen et al., 2022). The building with PV systems at roof saves 30%–50% energy as compared to roof with no installed PV (Paraschiv, Paraschiv, and Serban, 2021; Akhai, 2023 , Jia, S et al. 2024). Zero Energy Buildings comprise essential components like fenestration, walls, and roofs.

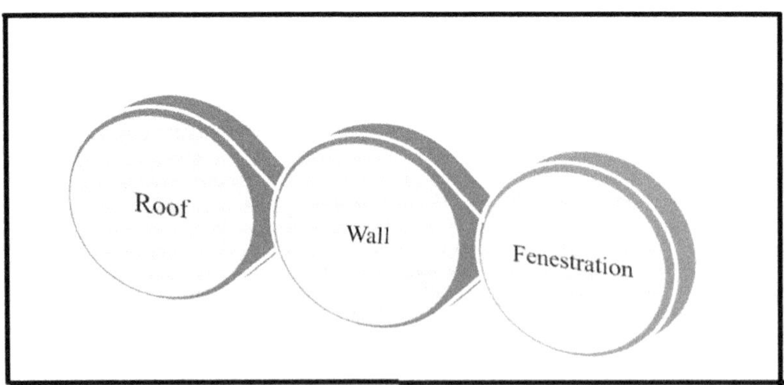

FIGURE 7.3 Elements of zero energy building.

7.3 METHODOLOGY

The pertinent terms pertaining to zero energy, high-performance buildings are studied to determine the characteristics that influence the thermal and energy performance of the building envelope. This includes components such as the wall system, the roof system, the fenestration, and the materials utilized on the building exterior. In order to provide designers, architects, engineers, and other stakeholders with assistance in choosing and developing high-performance components of the building envelope, the conclusions of case studies were gathered after relevant data from research papers, articles, and books were analyzed and assembled. The objective was to come up with different ways to save energy so that we could have sustainable structures.

7.4 FENESTRATION

By lowering lighting loads while maintaining the visual and thermal comfort for the building occupants, fenestrations impact the possibility of daylight harvesting. To lower the cooling load, the size of the building cooling systems, placement of windows, its size, and type of glazing should be strategically chosen. The visual and heat transmission properties of the glass and the direction and size of the opening must be precisely calibrated to achieve a balance between daylight penetration and heat gain (Hernández-Pérez et al., 2014; Omrany et al., 2016; Sarkar and Bose, 2016) and above all position of openings or apertures also affect the intensity of solar radiation like north-facing apertures receive less solar radiation as compared to south-facing facades. Openings (or walls) that face east and west get a lot of solar radiation all year long (Nematchoua et al. 2017; Cortês et al., 2021). The fenestrations (windows, skylights, and other openings in a building) do allow the solar radiation to let in but should let natural light and the prevailing wind into the structure (Akhai, 2023) for proper ventilation. However, significant heat gain may result from solar radiation passing through these fenestrations, particularly windows. Windows with glazing keep the heat inside the room. Sunlight's short-wave infrared radiation can easily flow through glass windows (Gao et al., 2023). If fenestration systems are not correctly constructed, the resultant temperature within the building may thus be even more significant than the outside temperature (Pan et al., 2023). Building fenestrations can significantly impact lighting and cooling loads from the standpoint of NZEB design. At the building's design stage, enhanced glazing can influence the embodied impact and assist in choosing the most environmentally friendly option. For example, using triple glazing rather than double glazing increases embodied energy percentage by 2 to 11 and reduces operational energy impact by 8.3% (Pan et al., 2023) whereas the frames such as polyvinyl chloride and fiber glass frames for triple glazing put less embodied impacts and better thermal behavior as compared to single glazing and double glazing. The architectural intervention, location, and orientation of the windows openings of commercial buildings in the United States can reduce the energy impact and improve the building's efficiency. They suggested that the energy consumption of buildings can be minimized if large windows are placed in the north direction and small windows on the south side (Pan et al., 2023). Even coatings used on fenestration can also contribute to building performance. The thermal performance of the building can be improved by using fenestration materials with anti-reflection properties and low emissivity (Akhai, Mala, and Jerin, 2021). In addition, if renewables are integrated, they will help to reflect infrared radiation emitted by the BIPV array, and the frame can thus enhance the energy performance of the building (Tanwar and Akhai, 2017). On the other hand, even though renewable energy systems are a good solution to climate change, carbon emissions during their manufacturing or whole life cycle should be taken into account for net-zero emery building (Akhai, Mala, and Jerin, 2020).

With the improvement in energy performance, there is a need to provide thermal and visual comfort. To achieve this, the authors optimized a smart shading device and window opening control methods to achieve the targets of NZEB, resulting in a 77% decrease in energy consumption without compromising thermal and visual comfort (Mathias, Juenger, and Horton, 2023). The authors proposed a two-story building design in Iran by optimizing the orientation, shading devices, window-to-wall ratio (WWR), glazing type, and the building to be designed as NZEB while installing insulating material like polyurethane or with low e glazing results into energy savings (Al-Mudhaffer et al., 2022). Modeling of an administrative building in Egypt's hot climate for skin parameters on the indoor daylighting and concluded that thermal energy efficiency could be achieved by adjusting the WWR following the path of the sun (Mehra et al., 2022; Prozuments et al., 2023). The above literature concludes that apart from the material used in windows, geometrical design factors, and window position, insulating material significantly affects building systems for heat gain and loss, thus affecting the building energy consumption. Typically, fenestration consumes 20% of embodied energy. Demonstration of high-performance fenestration like transparent surfaces or film can reduce thermal transmittance, and thus, operational energy can count down but suddenly shoots up embodied energy and embodied emissions (Pan et al., 2023). To improve the thermal performance of the building, one needs to focus on certain parameters like type of window glass, size of window, type of glazing, type of frame, and other elements such as thermal transmittance factor, visual transmittance, and solar heat gain coefficient as shown in Figure 7.4 which will help the designer to design the energy-efficient fenestration. Bronze reflecting glasses are worth more from a thermal and energy efficiency perspective when compared to single, double, triple, and quadruple glasses of clear, bronze, and green reflectivity (Gorantla et al., 2017).

7.5 ROOF TOP

As the environment is facing the most significant challenge due to climate change and in-built systems, the roof covering is constantly exposed to maximum heat, rainfall, wind effect, ultraviolet rays, and various other environmental factors that can result in the deterioration of materials. For NZEB design, interventions are to be done on roofs to reduce energy consumption and enhance thermal comfort. Considering the effect of roofing material on embodied energy, like concrete roof slab which is heat

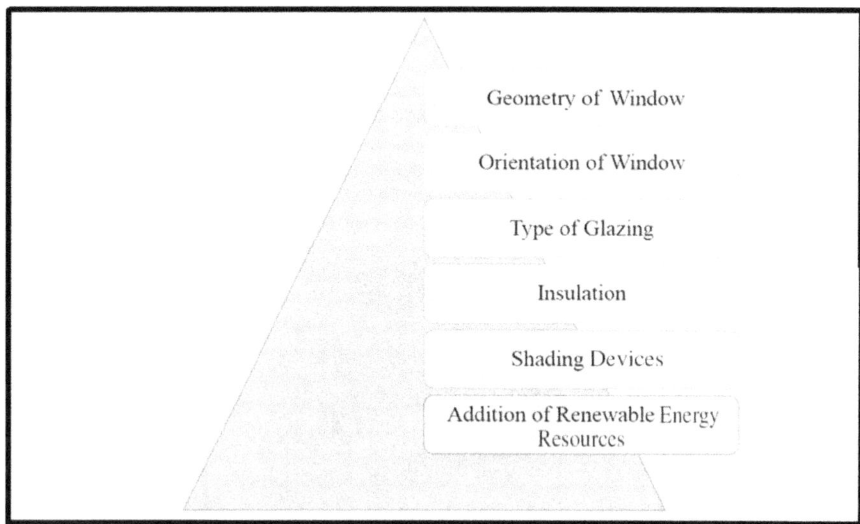

FIGURE 7.4 Parameters of fenestration.

sensitive and provides discomfort to the occupants beneath it, if retrofitted with cool tiles or terrazzo tiles can reduce the embodied emissions by 2–6% (Akhai, Bansal, and Singh, 2020; Kaushal, Singh, and Gupta, 2022). And if PV panels are implemented on roof top, the configuration of the proposed roof system can significantly improve energy efficiency and economic, social, and environmental benefits (Prabakaran, Lal, and Kim, 2023).

If PV systems are added to roofs which is not insulated this can result in energy savings of 55% in apartments whereas 80% in single family (Akhai, Singh, and John, 2016). The thickness of insulation plays a crucial role in heat gain and heat loss and the author found that 4 cm thickness of insulation out of 0, 2, 4, 6 cm is the optimized value in roof combinations for the existing residential building in Morocco (Habibi, Obonyo, and Memari, 2020). Considering the factors of temperature and color, the author investigated the effect of cool roofs and concluded that for tropical and sub-tropical regions the roofs with high reflectivity reduce total energy consumption by 14% and 22%, respectively whereas for cold and temperate regions the energy demand increases (Le, Whyte, and Biswas, 2019).

The authors analyzed the Italian residential building and proposed that cool clay tile as a cool roof solution can dip the temperature by 4.7 degrees Celsius, and highly reflective light-colored cool roofs are the best and most viable solution for the building (Huberman et al., 2015; Kaushal, 2022).

As roofs are a significant part of the building envelope, and in the case of old buildings, it increases the heat gain and heat loss, it requires more intervention against weathering action. The insulation is undoubtedly acting as the heat gain or loss barrier as shown in Figure 7.5. The authors proposed PV panels with roof insulation to improve reflectivity from the top roof which decreases the cooling and heating load to 39% and 28% which decreases the primary building energy consumption (Singh

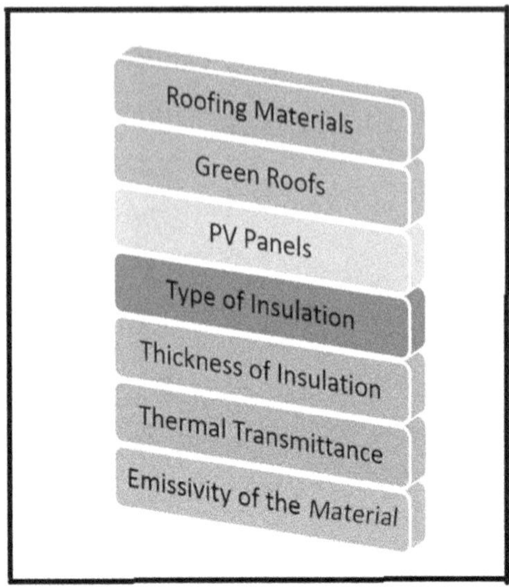

FIGURE 7.5 Parameters of roof.

and Chaudhary, 2021). To lower the temperature, the green roof is the other alternate solution reducing almost 70% of the cooling load. The authors investigated that green roofs reduce carbon emissions and can sequestrate carbon resulting in climate neutrality (Wagdy et al., 2015). The authors mentioned that vegetative roofs are superior to traditional gravel ballasted roofs (Klemm and Wiggins, 2016).

The above studies reflected that top roof is the most heat-absorbing component of the building envelope and contributes to 50% of thermal load (Husin and Harith, 2012). This heat transfer depends on the material used in roofing. Generally, the insulation is provided outside or inside the roofing to increase the thickness. Also, depending on the thermal properties of the insulating material like expanded polystyrene, a stone wool both have thermal transmittance of 0.34 w/M k and can act as a barrier to heat transfer, but on the other hand, if the embodied energy is compared expanded polystyrene consumes more embodied energy of 85 MJ as compared to stone wool, i.e., 25 MJ for its manufacturing. The material used in roofing may be concrete, steel, and addon insulating material are all highly energy-intensive. The above study shows that the provision of an addon can decrease the operational energy, but on the other hand, primary or embodied energy increases, which is to be optimized. For any building envelope, roof is the significant component in reducing the energy consumption; however, the reduction is done by thermal insulation of roof, and it is an essential part of that building envelope as it reduces the incoming heat flux (defined as the heat transfer per unit area per unit time) from or to the surface. Moreover, more than 60% of heat transfer occurs through the roof as the roof receives the largest amount of solar radiation per sq mt annually. Therefore different methods have been applied to reduce this solar energy gain and improve the energy performance of the building. If the additions are placed on the roof, they may result in an increase in the building's dead load; thus, the most practical alternative would be to maximize the thermal mass of the structure via the use of a life-cycle energy analysis and to determine its level of thermal performance.

7.6 WALL

Walls are the outer leaves of the building envelope that share the most prominent part of a building. They are the barriers between outdoor and indoor environments and are a route to heat transmission or heat loss. So, if the proper selection of the material or relevant module of the wall is constructed, it can reduce the energy consumption of the buildings. The author concluded that straw bales can reduce a wall's embodied energy and carbon emissions by about 7 tons compared to conventional walls (Nugrahanti, Lubis, and Kusyala, 2018).

According to the findings of the research (Azmy and Ashmawy, 2018), the embodied energy of a walling assembly may be decreased by 5,305 MJ for every thousand sun-dried bricks that are used rather than oven-dried bricks. The authors compared the green walls with conventional walls and found that total energy consumption and direct cooling demand reduces by 10.5% and 13% with green walls (Dabaieh et al., 2020; Kaushal et al., 2021). Therefore, the careful selection of wall modules can reduce energy levels. The authors explored the different modules of walls. Likewise, among Trombe Walls, AAC Block Walls, Cavity Walls, and Green Walls, it was found that green walls have shown the best results and can reduce building energy consumption and can act as insulators in buildings minimizing the solar radiations into the building (Feng et al., 2021). The modular living wall performs well in plant development and carbon sequestration and has a negative carbon balance of 22.7 kg CO_2/m^2 over its life-cycle period (Jelle, Kalnæs, and Gao, 2015; Akhai, 2023c). The authors estimated the walls of the building in a hot climate and estimated that insulation thickness also depends on the direction of the building; likewise, south, north, east, and west orientations as well as on climatic conditions (Norton et al., 2011; Kaushal, Gupta, and Bhowmick, 2017).

According to the literature mentioned above, heat gain and loss occur not only through the non-opaque part of the building but also the opaque walls which is almost 40% of the total heat. The author suggested that because the opaque and non-opaque building components are exposed to the outside environment; therefore, it is crucial to choose this building component correctly (Norton et al., 2011). Thus, a proper, careful, and trial selection of material can reduce the ample amount of energy consumption. Wall with the insulation is recommended for all climates. For more thermally comfortable space and less energy requirements, insulation of the walls reduces the conduction loss through all components of the building envelope.

The wall is the most important component of the structure, and it is also the section that allows the most heat to be transferred through it. Increasing the wall's thickness may reduce the amount of heat that is transferred, but the most effective strategy is to choose a material with a low embodied energy value and thermal conductivity as shown in Figure 7.6 that is optimized for the temperatures in the area.

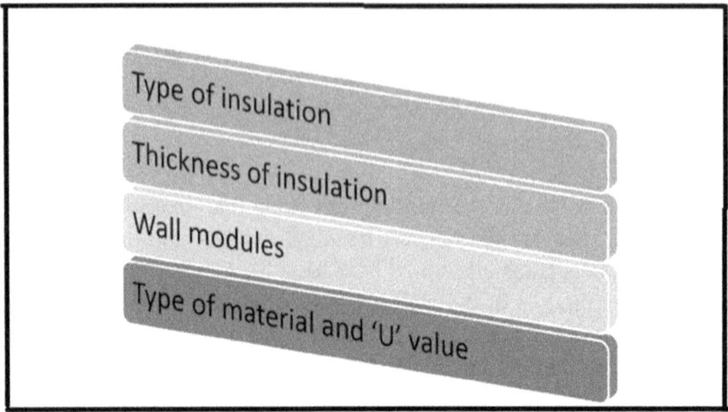

FIGURE 7.6 Parameters of wall.

7.7 MATERIAL USED ON THE ENVELOPE

Comparing AAC buildings to conventional buildings, energy consumption has increased by 47.83%. It may seem impossible to build a home out of mud, yet indigenous tribes all over the world have long understood the importance of clay-building materials. They have successfully mixed them with straw to build a shelter for thousands of years.

The author analyzed the effect of different materials like heavy materials, clay and insulation material and concluded that it is not only the material used in the building frame only matters but also the material used for insulating the building plays a role (Sayed et al., 2021). The author suggested that Ferro Cellular lightweight-concrete Insulated Panel Assembly for a building can replace brick masonry walls as strength parameters are equivalent to solid brick masonry and it is energy efficient and light in weight also (Rabani, Madessa, and Nord, 2021).

Analysis by the author showed that sustainable material can lead to a sustainable building envelope; likewise, limestone and marble, if used as cladding material or insulating material can reduce energy consumption and also lead to sustainability (Jelle, Kalnæs, and Gao, 2015), and for bricks it is seen that sun-dried bricks are much more energy efficient than fired clay bricks and can form energy-efficient building envelopes (ElBatran and Ismaeel, 2021).

The literature concluded that material selection is very significant and is key by which energy consumption can be reduced as cement, sand, brick, and steel account for 70% of embodied carbon and energy (ElBatran and Ismaeel, 2021). So, it is wise to reduce the quantity of the materials used by altering the technology or substituting the lower embodied energy materials like ground granulated blast slag, fly ash, etc. Overall literature reflects that every building combines raw and processed materials (Vijaykumar, Srinivasan, and Dhandapani, 2007), all contributing to the primary and building's total embodied energy. Optimizing the energy in terms of operational and embodied energy is viable to reduce the embodied energy in the building technology and processes. The author compared glass wool, rock wool, and EPS as insulation materials and concluded that insulating and other structural materials are vital to optimize the building envelope for high energy efficiency (Mokhtara et al., 2021; D'Agostino et al., 2022). As insulation on building envelopes could affect energy efficiency and it is the fundamental key and the author found that EPS insulation if used on the exterior wall (Abdou et al., 2021; Rawat and Singh, 2022) will be more effective as compared to any other type of insulation and concluded that careful selection of insulating material could be beneficial to reducing operation energy and improving the environmental scenario same as said by the authors Grazieschi, Asdrubali, and Thomas (2021) that aerogels and vacuum insulating panels liberate 11–19 kg CO_2eq/FU as comparative to traditional inorganic insulating materials glass wool release 16–31 kg CO_2eq/FU. Insulating materials shown in Figure 7.7 are undoubtedly the solution to reducing operational energy, which improves the profile. The operational energy overpowered the embodied energy by providing insulation in new and old buildings. Still, on the other hand, these materials exhibit embodied energy which affects the environment profile.

A material which is most appropriate for the construction technique is those material which have low energy cost considering all energy consumption during the life cycle. For this reason, when a selection of the material is made considering the energy efficiency, it is required to take the energy consumption and its emissions at all stages including extraction of the material from the source and its transportation, its uses, its destruction, and recycling as a whole. Thus, energy-efficient material decreases the fossil-based energy resource and makes contribution to overcome

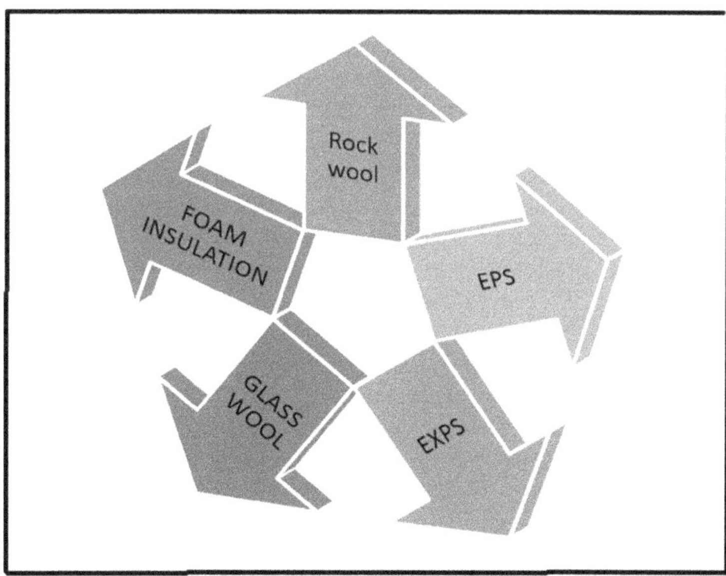

FIGURE 7.7 Types of insulation used in zero energy buildings.

environmental degradation. Not only should the mechanical, physical, and chemical qualities of the material that is going to be incorporated into the structure be analyzed, but it should also be thermally tested, such as with AAC blocks, fly ash bricks, and other construction materials.

7.8 FEASIBILITY OF CONSTRUCTING BUILDING ENVELOPE FOR NZEB

In recent years, there has been a growing interest in constructing buildings that have a minimal environmental impact. One way to achieve this is by designing and constructing buildings that produce as much energy as they consume. These buildings are known as NZEBs.

- One of the key components of an NZEB is the building envelope, which consists of the walls, roof, and windows. The author presented the potential of Trombe wall and suggested that although this wall has great potential in reducing the heating need in cold regions, in addition to this if supplemented with PCM/Building Integrated Photo Voltaic, this whole system can reduce the space heating needs to 10% (Kaushal, Gupta, and Bhowmick, 2018; Pisello, Cotana, and Brinchi, 2014). So, this shows that the building envelope plays a critical role in reducing energy consumption and environmental emissions. By using high-performance building envelopes and facades, as well as opaque and non-opaque features, the energy required to heat and cool the building can be significantly reduced.
- Another important aspect of NZEBs is the embodied energy of the building materials. Embodied energy refers to the energy consumed during the entire life cycle of a building material, including the extraction, manufacturing, transportation, installation, use, and disposal phases evaluated the different insulating material for cooling effect and their relative GWP by life-cycle assessment and resulted that out of extruded polystyrene, expanded polystyrene, rock wool, and glass wool and cellulose, extruded polystyrene has high cellulose and has the least GWP (Kaushal, Gupta, and Bhowmick, 2021). By using sustainable and high-energy-intensive materials, the total embodied energy of the building can be reduced as a result that prefabricated materials can lead to a decrease in the building's overall environmental impact (Bhuvad, 2022; Kaushal, Gupta, and Bhowmick, 2019).
- However, it is important to note that while the operational energy consumption of NZEBs is significantly reduced, the embodied energy contribution increases. This means that careful selection of materials with the right combination of thermal properties and embodied energy is crucial to achieving a sustainable and true NZEB.

Thus, constructing buildings with a minimal environmental impact is not only desirable but also achievable. By focusing on the building envelope and using sustainable and high-energy-intensive materials, NZEBs can significantly reduce energy consumption and environmental emissions, contributing to a more sustainable future.

7.9 CONCLUSION

This chapter explained the feasibility of constructing the building envelope for NZEB. Energy consumption associated with any country's environmental emissions in the buildings can also be reduced by constructing the NZEB. Today the main focus of modern buildings or NZEB is to reduce the operational energy consumption, which decreases the total energy consumption but augments the embodied energy addon as insulation to the roof and walling system that reduces the operational power. On the other side, there is an enormous jump in embodied energy as reflected by EPS, which sounds like a good insulator, but its emissions and embodied energy are much higher than glass wool. High-performance building envelopes, facades, or opaque and non-opaque features (roof, wall, fenestration, insulation, thermal properties of the material chosen) are crucial components in the design of buildings for their transformation as the NZEB. The contribution of embodied energy appears to increase from 20% to 60% toward NZEB, which was 15–20% in conventional buildings, even though there is an essential reduction of 40–50% in total energy consumption by modifying the building envelope with sustainable and high-energy-intensive material. Although designing and selecting parameters is necessary for constructing a successful NZEB, the careful selection of the material with an association of thermal properties and embodied energy is most important. The variation of embodied energy indicated by the whole life-cycle energy analysis for its selective system boundaries must be considered in energy efficiency regulation and for future standardization. This direction will achieve sustainable and true NZEB to achieve a circular economy, and technology should be considered to control the environmental profile. The components of the building envelope are the main source of heat loss or heat gain so choice of the materials and the module making of these components can help to reduce the energy consumption as well as carbon emissions.

7.10 LIMITATIONS

There has not been much research into the useful lives of the building's components, which might be helpful in providing data that is more distinctive and supplementary. It is necessary to evaluate not only the behavior of the building during the peak load time but also the shape and form of the building, as well as the behavior of the structure in relation to its orientation.

7.11 FUTURE SCOPE

While the research on building envelope components for intelligent civil infrastructure has made significant strides, there are still areas where further investigation is needed.

- First, the research on the thermal properties of different building materials and components with embodied energy should be further optimized. The goal is to reduce the embodied energy while still maintaining a high level of energy efficiency.
- Second, more studies are needed on the life-cycle analysis of the building envelope components, including the materials used in their production and their end-of-life disposal. This analysis can provide a more comprehensive understanding of the environmental impact of these components and help in the development of more sustainable alternatives.
- Third, there is a need to explore the integration of renewable energy sources into the building envelope, such as photovoltaic panels and solar thermal collectors. This integration could increase the energy efficiency of the building envelope and contribute toward the goal of achieving net-zero energy buildings.
- Finally, it is essential to develop standards and regulations for energy-efficient buildings that incorporate the latest research and technologies like phase change materials, intelligent structural health monitoring systems, thermal break technology, dynamic facades, and so on.
- These standards can guide designers, builders, and policymakers toward a more sustainable future for the construction industry.

Thus, future research on building envelope components for intelligent civil infrastructure should focus on the optimization of thermal properties, life-cycle analysis, integration of renewable energy sources, and the development of standards and regulations. These efforts will help us achieve a more sustainable and energy-efficient built environment.

7.12 CHALLENGES

- As the need for more responsible use of resources and energy grows more critical, the building and construction sector has been making considerable efforts toward more sustainable building practices. However, one of the most difficult issues that both builders and architects must overcome is determining how well their structures perform in terms of thermal efficiency and energy consumption. This method often includes the use of metrics like R-value, U-value, and thermal transmittance, which are used to quantify the insulating characteristics of a structure as well as its capacity to either retain or release heat.
- While these measures can give useful information, they also have the potential to create substantial challenges for the design of buildings. This is especially important to keep in mind whether one is seeking to include active or passive tactics in the design of a structure. Active strategies are those that must have an input of energy in order to function, such as a building's heating and cooling systems. Passive strategies, on the other hand, depend on naturally occurring processes to control the inside temperature of a structure, such as solar gain and ventilation.

- When analyzing the performance of buildings using these measures, one of the most significant problems that often arises is that these metrics frequently fail to take into account the whole of the building's life cycle, beginning with its conception and ending with its demolition. When determining the energy efficiency of a building, it is common practice to ignore the embodied energy, which refers to the amount of energy needed to produce and transport all of the components that go into the construction of the building. In addition, an energy flow study has to be carried out in order to have a comprehensive understanding of how energy is being used inside the structure and where it could be squandered.
- The selection of the appropriate building type, as well as the integration of both opaque and non-opaque elements within the structure, is still another essential aspect of architectural design. These options are quite sensitive to the weather patterns that prevail in the region, and as a result, they need to be regulated properly. For instance, structures located in regions with lower average temperatures may need increased thickness of insulation and fewer windows, while structures located in climates with higher average temperatures may prioritize natural ventilation to minimize the need for artificial cooling (Akhai S 2023b)
- It is necessary for architects and builders to have access to a complete structural database that gives benchmarking data related to thermal properties and embodied energy in order for them to be able to make educated choices on the design of buildings. A database of this kind does not seem to be available currently. Therefore, to assist the development of sustainable construction practices, there is a need for more study as well as the gathering of data in this field.

Thus, it is vital for the design of sustainable buildings to include an assessment of the performance of the building using thermal and energy metrics. The intricacy of the components that are involved in this process, such as embodied energy and energy flow analysis, might make it difficult to complete this procedure successfully. In addition, the decisions that are taken about the kind of structure and the selection of its components need to be adapted to the weather patterns of the area. A complete structural database that takes into consideration these elements is required to facilitate the making of educated choices about the design of buildings.

REFERENCES

Abdou, N., Mghouchi, Y.E., Hamdaoui, S., Asri, N.E. and Mouqallid, M., 2021. Multi-objective optimization of passive energy efficiency measures for net-zero energy building in Morocco. *Building and environment*, *204*, p.108141.

Aggarwal, P., Rana, M. and Akhai, S., 2022. Briefings on e-waste hazard until COVID era in India. *Materials Today: Proceedings*, *71*, pp.389-393.

Akhai, S., 2023a. Biogas Plants for Sustainable Municipal Waste Management: A Brief Review. *Handbook of Research on Safe Disposal Methods of Municipal Solid Wastes for a Sustainable Environment*, pp.162–179.

Akhai, S., 2023b. From black boxes to transparent machines: The quest for explainable AI. Available at SSRN: https://ssrn.com/abstract=4390887 or http://dx.doi.org/10.2139/ssrn.4390887.

Akhai, S., 2023c. Navigating the Potential Applications and Challenges of Intelligent and Sustainable Manufacturing for a Greener Future.

Akhai, S., Bansal, S.A. and Singh, S., 2020. A critical review of thermal insulators from natural materials for energy saving in buildings. *Journal of Critical Reviews*, *7*(19), pp.278–283.

Akhai, S. and Khang, A., 2024. Energy Efficiency and Human Comfort: AI and IoT Integration in Hospital HVAC Systems. *Medical Robotics and AI-Assisted Diagnostics for a High-Tech Healthcare Industry*, pp.93–108.

Akhai, S., Mala, S. and Jerin, A.A., 2020. Apprehending air conditioning systems in context to COVID-19 and human health: a brief communication. *International Journal of Healthcare Education & Medical Informatics (ISSN: 2455-9199)*, *7*(1&2), pp.28–30.

Akhai, S., Mala, S. and Jerin, A.A., 2021. Understanding whether air filtration from air conditioners reduces the probability of virus transmission in the environment. *Journal of Advanced Research in Medical Science & Technology (ISSN: 2394-6539)*, *8*(1), pp.36–41.

Akhai, S., Singh, V.P. and John, S., 2016. Investigating Indoor Air Quality for the Split-Type Air Conditioners in an Office Environment and Its Effect on Human Performance. *Journal of Mechanical Civil Engineering*, *13*(6), pp.113–118

Akhai, S., Thareja, P. and Singh, V.P., 2017. Assessment of Indoor Environment Health Sustenance in Air Conditioned Class Rooms. *Advanced Research in Civil and Environmental Engineering*, *4*(1&2), pp.1–9.

Alam, M.A., Kumar, R., Yadav, A.S., Arya, R.K. and Singh, V.P., 2023. Recent developments trends in HVAC (a, ventilation, and air-conditioning) systems: A comprehensive review. *Materials today: proceedings.*

Al-Mudhaffer, A.F., Saleh, S.K. and Kadhum, G.I., 2022. The role of sustainable materials in reducing building temperature. *Materials Today: Proceedings*, *61*, pp.690–694.

Asdrubali, F., Grazieschi, G., Roncone, M., Thiebat, F. and Carbonaro, C., 2023. Sustainability of building materials: Embodied energy and embodied carbon of masonry. *Energies*, *16*(4), p.1846.

Azmy, N.Y. and Ashmawy, R.E., 2018. Effect of the window position in the building envelope on energy consumption. *International Journal of Engineering & Technology*, *7*(3), pp.1861–1868.

Bhuvad, S.S., 2022. Investigation of annual performance of a building shaded by rooftop PV panels in different climate zones of India. *Renewable Energy*, *189*, pp.1337–1357.

Chen, W., Yang, S., Zhang, X., Jordan, N.D. and Huang, J., 2022. Embodied energy and carbon emissions of building materials in China. *Building and environment*, *207*, p.108434.

Christopher, S., Vikram, M.P., Bakli, C., Thakur, A.K., Ma, Y., Ma, Z., Xu, H., Cuce, P.M., Cuce, E. and Singh, P., 2023. Renewable energy potential towards attainment of net-zero energy buildings status–a critical review. *Journal of Cleaner Production*, p.136942.

Cortês, A., Almeida, J., Santos, M.I., Tadeu, A., de Brito, J. and Silva, C.M., 2021. Environmental performance of a cork-based modular living wall from a life-cycle perspective. *Building and Environment*, *191*, p.107614.

Dabaieh, M., Heinonen, J., El-Mahdy, D. and Hassan, D.M., 2020. A comparative study of life cycle carbon emissions and embodied energy between sun-dried bricks and fired clay bricks. *Journal of Cleaner Production*, *275*, p.122998.

D'Agostino, D., Parker, D., Melià, P. and Dotelli, G., 2022. Optimizing photovoltaic electric generation and roof insulation in existing residential buildings. *Energy and Buildings, 255*, p.111652.

ElBatran, R.M. and Ismaeel, W.S., 2021. Applying a parametric design approach for optimizing daylighting and visual comfort in office buildings. *Ain Shams Engineering Journal, 12*(3), pp.3275–3284.

Feng, F., Kunwar, N., Cetin, K. and O'Neill, Z., 2021. A critical review of fenestration/window system design methods for high performance buildings. *Energy and Buildings, 248*, p.111184.

Füchsl, S., Rheude, F. and Röder, H., 2022. Life cycle assessment (LCA) of thermal insulation materials: A critical review. *Cleaner Materials, 5*, p.100119.

Gao, Z., Liu, H., Xu, X., Xiahou, X., Cui, P. and Mao, P., 2023. Research progress on carbon emissions of public buildings: a visual analysis and review. *Buildings, 13*(3), p.677.

Gorantla, K., Shaik, S. and Setty, A.B.T.P., 2017. Effects of single, double, triple and quadruple window glazing of various glass materials on heat gain in green energy buildings. In *Materials, Energy and Environment Engineering: Select Proceedings of ICACE 2015* (pp. 45–50). Springer Singapore.

Goyal, R., 2023. Smart Infrastructure for Smarter Transportation: Innovations in Traffic Management and Adaptive Road Systems. *Journal of Advanced Research in Automotive Technology and Transportation System, 7*(2), pp.5_10–5_10.

Goyal, R. and Dhanoa, G.S., 2022. Advancement in Energy Efficiency and Carbon Reduction in Buildings for Sustainable Future.

Grazieschi, G., Asdrubali, F. and Thomas, G., 2021. Embodied energy and carbon of building insulating materials: A critical review. *Cleaner Environmental Systems, 2*, p.100032.

Habibi, S., Obonyo, E.A. and Memari, A.M., 2020. Design and development of energy efficient re-roofing solutions. *Renewable Energy, 151*, pp.1209–1219.

Hernández-Pérez, I., Álvarez, G., Xamán, J., Zavala-Guillén, I., Arce, J. and Simá, E., 2014. Thermal performance of reflective materials applied to exterior building components—A review. *Energy and Buildings, 80*, pp.81–105.

Huberman, N., Pearlmutter, D., Gal, E. and Meir, I.A., 2015. Optimizing structural roof form for life-cycle energy efficiency. *Energy and Buildings, 104*, pp.336–349.

Husin, S.N.F.S. and Harith, Z.Y.H., 2012. The performance of daylight through various type of fenestration in residential building. *Procedia-Social and Behavioral Sciences, 36*, pp.196–203.

Jelle, B.P., Kalnæs, S.E. and Gao, T., 2015. Low-emissivity materials for building applications: A state-of-the-art review and future research perspectives. *Energy and Buildings, 96*, pp.329–356.

Kaushal, S., 2022. Microstructure and tribological characterization of composite castings developed through In-situ microwave hybrid heating. *International Journal of Metalcasting, 16*(4), pp.2150–2161.

Kaushal, S., Gupta, D. and Bhowmick, H., 2017. Investigation of dry sliding wear behavior of Ni–SiC microwave cladding. *Journal of Tribology, 139*(4), p.041603.

Kaushal, S., Gupta, D. and Bhowmick, H., 2018. On processing of Ni-Cr3C2 based functionally graded clads through microwave heating. *Materials Research Express, 5*(6), p.066405.

Kaushal, S., Gupta, D. and Bhowmick, H., 2019. On processing and flexural behaviour of functionally graded clads developed through microwave irradiation. *Materials Research Express, 6*(7), p.076405.

Kaushal, S., Gupta, D. and Bhowmick, H., 2021. Wear behavior of microwave-processed Ni-WC8Co-based functionally graded materials. *Proceedings of the Institution of*

Mechanical Engineers, Part L: Journal of Materials: Design and Applications, *235*(5), pp.1036–1045.

Kaushal, S., Singh, S. and Gupta, D., 2022. Processing strategy for high strength Ni-based hybrid composite clad on SS 316L steel through microwave heating. *Proceedings of the Institution of Mechanical Engineers, Part B: Journal of Engineering Manufacture*, *236*(3), pp.190–203.

Kaushal, S., Singh, D., Gupta, D. and Jain, V., 2019. Wear resistance improvement of austenitic 316 L steel by microwave-processed composite clads. *Journal of Tribology*, *141*(4), p.041605.

Khan, M.Y. and Baqi, A., 2021. Experimental and theoretical analysis of a new kind of building envelope. *Materials Today: Proceedings*, *43*, pp.1368–1375.

Klemm, A. and Wiggins, D., 2016. Sustainability of natural stone as a construction material. In *Sustainability of construction materials* (pp. 283–308). Woodhead Publishing.

Koura, J., Manneh, R., Belarbi, R., El Khoury, V. and El Bachawati, M., 2020. Comparative cradle to grave environmental life cycle assessment of traditional and extensive vegetative roofs: An application for the Lebanese context. *The International Journal of Life Cycle Assessment*, *25*, pp.423–442.

Kumar, P. and Akhai, S., 2022. Effective energy management in smart buildings using VRV/VRF systems. In *Additive Manufacturing in Industry 4.0* (pp. 27–35). CRC Press.

Le, A.B.D., Whyte, A. and Biswas, W.K., 2019. Carbon footprint and embodied energy assessment of roof-covering materials. *Clean Technologies and Environmental Policy*, *21*, pp.1913–1923.

Li, X. and Tingley, D.D., 2023. A whole life, national approach to optimize the thickness of wall insulation. *Renewable and Sustainable Energy Reviews*, *174*, p.113137.

Lu, Y., Cui, P. and Li, D., 2016. Carbon emissions and policies in China's building and construction industry: Evidence from 1994 to 2012. *Building and Environment*, *95*, pp.94–103.

Mathias, J.A., Juenger, K.M. and Horton, J.J., 2023. Advances in the energy efficiency of residential appliances in the US: A review. *Energy Efficiency*, *16*(5), p.34.

Mehra, S., Singh, M., Sharma, G., Kumar, S., Navishi and Chadha, P., 2022. Impact of construction material on environment. *Ecological and Health Effects of Building Materials*, pp.427–442.

Mihalakakou, G., Souliotis, M., Papadaki, M., Menounou, P., Dimopoulos, P., Kolokotsa, D., Paravantis, J.A., Tsangrassoulis, A., Panaras, G., Giannakopoulos, E. and Papaefthimiou, S., 2023. Green roofs as a nature-based solution for improving urban sustainability: Progress and perspectives. *Renewable and Sustainable Energy Reviews*, *180*, p.113306.

Mneimneh, F., Ghazzawi, H. and Ramakrishna, S., 2023. Review study of energy efficiency measures in favor of reducing carbon footprint of electricity and power, buildings, and transportation. *Circular Economy and Sustainability*, *3*(1), pp.447–474.

Mokhtara, C., Negrou, B., Settou, N., Bouferrouk, A. and Yao, Y., 2021. Optimal design of grid-connected rooftop PV systems: An overview and a new approach with application to educational buildings in arid climates. *Sustainable Energy Technologies and Assessments*, *47*, p.101468.

Nematchoua, M.K., Ricciardi, P., Reiter, S. and Yvon, A., 2017. A comparative study on optimum insulation thickness of walls and energy savings in equatorial and tropical climate. *International Journal of Sustainable Built Environment*, *6*(1), pp.170–182.

Norton, B., Eames, P.C., Mallick, T.K., Huang, M.J., McCormack, S.J., Mondol, J.D. and Yohanis, Y.G., 2011. Enhancing the performance of building integrated photovoltaics. *Solar Energy*, *85*(8), pp.1629–1664.

Nugrahanti, F.I., Lubis, I.H. and Kusyala, D., 2018, December. The impact of building mass configuration towards wind-driven natural ventilation in apartment in Jakarta. In *IOP Conference Series: Earth and Environmental Science* (Vol. 213, No. 1, p. 012042). IOP Publishing.

Omrany, H., Ghaffarianhoseini, A., Ghaffarianhoseini, A., Raahemifar, K. and Tookey, J., 2016. Application of passive wall systems for improving the energy efficiency in buildings: A comprehensive review. *Renewable and sustainable energy reviews*, *62*, pp.1252–1269.

Pan, Y., Zhu, M., Lv, Y., Yang, Y., Liang, Y., Yin, R., Yang, Y., Jia, X., Wang, X., Zeng, F. and Huang, S., 2023. Building energy simulation and its application for building performance optimization: A review of methods, tools, and case studies. *Advances in Applied Energy*, p.100135.

Paraschiv, S., Paraschiv, L.S. and Serban, A., 2021. Increasing the energy efficiency of a building by thermal insulation to reduce the thermal load of the micro-combined cooling, heating and power system. *Energy Reports*, *7*, pp.286–298.

Pisello, A.L., Cotana, F. and Brinchi, L., 2014. On a cool coating for roof clay tiles: development of the prototype and thermal-energy assessment. *Energy procedia*, *45*, pp.453–462.

Prabakaran, R., Lal, D.M. and Kim, S.C., 2023. A state of art review on future low global warming potential refrigerants and performance augmentation methods for vapour compression based mobile air conditioning system. *Journal of Thermal Analysis and Calorimetry*, *148*(2), pp.417–449.

Pragati, S., Shanthi Priya, R., Pradeepa, C. and Senthil, R., 2023. Simulation of the energy performance of a building with green roofs and green walls in a tropical climate. *Sustainability*, *15*(3), p.2006.

Prozuments, A., Borodinecs, A., Bebre, G. and Bajare, D., 2023. A review on Trombe wall technology feasibility and applications. *Sustainability*, *15*(5), p.3914.

Rabani, M., Madessa, H.B. and Nord, N., 2021. Achieving zero-energy building performance with thermal and visual comfort enhancement through optimization of fenestration, envelope, shading device, and energy supply system. *Sustainable Energy Technologies and Assessments*, *44*, p.101020.

Rawat, M. and Singh, R.N., 2022. A study on the comparative review of cool roof thermal performance in various regions. *Energy and Built Environment*, *3*(3), pp.327–347.

Sarkar, A. and Bose, S., 2016. Exploring impact of opaque building envelope components on thermal and energy performance of houses in lower western Himalayans for optimal selection. *Journal of Building Engineering*, *7*, pp.170–182.

Sayed, E.T., Wilberforce, T., Elsaid, K., Rabaia, M.K.H., Abdelkareem, M.A., Chae, K.J. and Olabi, A.G., 2021. A critical review on environmental impacts of renewable energy systems and mitigation strategies: Wind, hydro, biomass and geothermal. *Science of the total environment*, *766*, p.144505.

Singh, D. and Chaudhary, R., 2021. Impact of roof attached Photovoltaic modules on building material performance. *Materials Today: Proceedings*, *46*, pp.445–450.

Sodagar, B., Rai, D., Jones, B., Wihan, J. and Fieldson, R., 2011. The carbon-reduction potential of straw-bale housing. *Building Research & Information*, *39*(1), pp.51–65.

Song, Y., Zhang, H. and Mithraratne, N., 2023. Research on influences of wall design on embodied and operating energy consumption: A case study of temporary building in China. *Energy and Buildings*, *278*, p.112628.

Tanwar, N. and Akhai, S., 2017. Survey Analysis for Quality Control Comfort Management in Air Conditioned Classroom. *Journal of Advanced Research in Civil and Environmental Engineering*, *4*(1&2), pp.20–23.

Tavares, V., Soares, N., Raposo, N., Marques, P. and Freire, F., 2021. Prefabricated versus conventional construction: Comparing life-cycle impacts of alternative structural materials. *Journal of Building Engineering*, *41*, p.102705.

Vijaykumar, K.C.K., Srinivasan, P.S.S. and Dhandapani, S., 2007. A performance of hollow clay tile (HCT) laid reinforced cement concrete (RCC) roof for tropical summer climates. *Energy and Buildings*, *39*(8), pp.886–892.

Wagdy, A., Elghazi, Y., Abdalwahab, S. and Hassan, A., 2015. The balance between daylighting and thermal performance based on exploiting the kaleidocycle typology in hot arid climate of Aswan, Egypt. In *AEI 2015* (pp. 300–315).

Jia, S., Weng, Q., Yoo, C. *et al.* Building energy savings by green roofs and cool roofs in current and future climates. *npj Urban Sustain* **4**, 23 (2024).

8 Examining Barriers for Implementation of Green Manufacturing in Indian MSMEs by Means of AHP Technique

Charanjit Singh, Davinder Singh, and Sukhjinder Singh

8.1 INTRODUCTION

Green manufacturing (GM) is the regeneration of production methods and the formation of procedures in an organization that does not affect the environment. With the aim of enhancing organizational performance, green manufacturing concepts provide management strategies for manufacturing companies (Singh et al. 2018). It refers to the manufacturing of environment-friendly products, mainly used in renewable energy systems. Manufacturing organizations can become "green organizations" by reducing waste and pollution through minimal use of natural resources and adopting reduce, recycle, and reuse policy. The ways by which manufacturing companies can go green are by conducting an energy audit and the integration of green principles, in particular, emphasizing the ability to handle the bottom line, or the environmental and social aspects of an industrial unit. Green manufacturing seems to be a big step in the direction of increasing an organization's environmental efficiency (Singh et al. 2021).

8.1.1 Green Manufacturing Objectives

The main objective of GM is to reduce and/or abolish the use of hazardous materials in the design, production, and application of various products or processes in order to avoid pollution and save natural resources by utilizing new knowledge. GM is considered as a technique that is capable of providing profit to an organization through operating methods that are not harmful to the environment. Recent times have also changed the customer perspective as nature lovers are intended to buy products that

DOI: 10.1201/9781003449225-8

have been produced by using methods/techniques that protect the environment. These types of customers are known as "Green Consumers," and they only buy environmentally friendly products (Young et al. 1997).

Reducing harmful waste shall improve the green image (Mohanty et al. 2014. MSMEs should emphasize on treatment of waste in order to minimize the effect of this waste on environment and at the same time maximizing the efficiency of present resources. There are a number of methods that can be used to improve productivity of an organization without affecting the natural environment; for example, design modifications, utilization of clean energy, and implementation of new/modified processes (Hicks and Dietmar 2007).

There are a number of environmental practices that can be used by various manufacturing organizations, but a few research studies are available that tried to explain the inspirations behind the selection of these practices as far as the viewpoint of MSMEs is concerned. There can be a large number of barriers that prevent MSMEs from engaging in environmental practices. Manufacturing organizations especially MSMEs are feeling the pressure due to government/policy issues which force them to maintain their manufacturing processes without harming the environment. Green business is derived from various factors such as customer needs, pressure from competitors, and effective utilization of natural resources (Baines et al. 2012). In densely populated countries like India, it is the need of the hour to reduce energy consumption, reduce water wastage, proper treatment of hazardous materials, reduce pollution, and adopt green purchasing. Manufacturing organizations that are adopting GM are gaining competitive advantages as compared with other manufacturing firms that are non-green manufacturing. In order to attain real benefits, Indian MSMEs need to follow certain fundamental practices; for example, they should commit to implementing a green philosophy and try to include GM operations in their products, processes, and routine operations (Hutchinson and Chaston 1994).

The barriers for green manufacturing practices in Malaysian SMEs have affected the manufacturing performance (Ghazilla et al. 2015). It is now the area of interest how manufacturing industries in developing nations, especially in agriculture sector, contribute to the growth and development of the country. At the same time, large quantity of waste produced by manufacturing activities is also a matter of concern for nations across the globe. This is because effective waste treatment or environmental protection activities have not been implemented properly in these industries. Manufacturing industries including food processing, mineral exploitation, petroleum, and the textile industry have historically released harmful materials in Ghana (Agan et al. 2013).

Although green methods have been tried in many other sectors, including manufacturing, services, construction, hospitals and so on, the adoption rate is still quite low in underdeveloped nations. As a result, GM implementation challenges are more severe in underdeveloped nations. Green assessment is a further action meant to stop the damaging effects of industrial activity on the environment. A set of nine steps that includes scoping actions, background research, impact evaluation, mitigation measures, comparing alternatives, records keeping, making decisions, and post audits are necessary for effective GM implementation. A successful application of these recommendations aids in preventing or reducing the detrimental impacts on the environment caused due to manufacturing activities. Another method to lessen the harm that industrial pollution causes to the environment is the deployment of preventive techniques. It includes redesigning equipment, upgrading products, altering processes, and recovering waste from manufacturing for recycling. Nevertheless, these activities are merely being done on a small, individual, or business scale. Due to some businesses' outdated technology, a number of these guidelines are no longer applicable (Castka et al. 2009; Singh et al. 2021).

The environmental sustainability of the Mexican maquiladora sector is significantly impacted by employee involvement and training. GM techniques go well beyond compliance as these are a way of life in the manufacturing industry that also has long-term positive effects on organizations' sustainability, success, and efficiency. Green tomorrow can be obtained by optimizing the energy utilized in manufacturing operations, enhanced use of renewable resources, waste reduction, and employing an incessant improvement strategy to minimize energy consumption (Daily et al. 2012).

8.1.2 Micro, Small, and Medium Enterprises (MSMEs)

MSMEs are regarded as the backbone of any nation because of their mammoth involvement in developing recent technologies, generating employment opportunities, and covering a large proportion of the modern economy. These units comprise almost 90% of the total industrial units in almost all the developing nations thereby generating a large amount of employment for rural as well as urban regions. Their share in industrial production is also accountable. MSME sector in India plays an important role as it covers major portion of the overall economy. When comparing the MSME sector to the entire industrial sector of any nation, this sector has continuously shown faster development in the past few decade (Hutchinson and Chastonet al. 1994) (Hutchinson et al., 1994).

8.1.3 The Indian Scenario

The MSME sector has grown to be a thriving and dynamic area of the economy. The Indian economy can overtake the United States as the second largest in the world by 2050 and become the third largest after China and the United States by 2032 with annual growth rates of over 8% till 2020 (Ministry of MSMEs). This sector has a well-established role in the economic and social growth of the nation. Despite a slow and unstable economy, MSMEs emerge as leaders throughout the recession, recovering jobs and commercial activity lost during the period. MSME sector adds almost 40% of the total manufacturing production in the country. Growth rates in this sector have been particularly excellent during the recent past. The total number of small industrial units has expanded from an estimated 0.87 million in 1982–83 to more than 3 million this year. The performance of MSME sector as compared with manufacturing sector as a whole imparts confidence that is essential for implementing GM in these industries (Singh et al. 2014).

In a study conducted by Ferreira et al. (2023), a theoretical framework was created to evaluate the effect of Green Supply Chain Management performance on the adoption of wire and arc additive manufacturing (WAAM) technology. Instead of creating new items, WAAM can be used to restore and repair old ones, extending their lifespan. These two factors can help a business save a lot of money by, for example, eliminating the need to store replacement parts. In direction to green manufacturing to cut carbon emissions and boost green production, some renowned multinational auto manufacturing companies have recently announced that they would stop producing automobiles with internal combustion engines. For instance, Volvo has stated that all new vehicles would be totally electric by 2030, Audi aims to stop manufacturing gasoline-powered vehicles in 2033, Honda will do so in 2040, and BYD will stop manufacturing gasoline-powered vehicles and will only create alternative energy vehicles starting in March 2022. However, there are still a number of challenging problems with the development of alternative energy cars, such as battery life, prolonged charge periods, a lack of charging infrastructure, and consumer reluctance. Alternative fuel vehicles probably won't be fully adopted for a very long time (Zhu et al. 2023).

There are a few studies conducted on GM practices to be implemented in Indian MSMEs that provide an overview of the concept. There are presently about 11 million MSME businesses in India, producing over 8000 items and contributing about 35% of the industrial exports. There are various online business directories and commerce portals that have a large database of manufacturers, suppliers, and customers. Also, less capital is needed to start a small-scale unit and maintain its day-to-day activities. Government banks and other financial institutes are willing to provide funds to these MSME units because of their growing value. This sector has become the largest contributor in terms of economy and production capacity of the country. Despite providing enormous contributions to nation's economy and having great opportunities to grow in the coming future, this sector will have to face certain challenges related to environmental concerns (Singh and Thakar 2018). Therefore, this study identifies and analyses the barriers to implementing GM in Indian MSMEs which will be beneficial for manufacturers and policymakers to get the desired outcomes.

The chapter is organized in four sections. Section 8.2 covers literature review for GM barriers. Section 8.3 provides the methodology for study and covers Pareto analysis and AHP approach. Section 8.4 discusses step-by-step implication of AHP approach. Conclusions of the study are summarized in Section 8.5.

8.2 LITERATURE REVIEW

Xu Jie (2017) describes that GM will be a widely used concept in manufacturing industries future because of its important contribution to preventing harmful effects on the environment and also preserving natural resources. This study explains the fundamental analysis of various techniques that are employed to effectively implement GM in manufacturing industries thereby protecting the environment and valuable resources for future generations. Kothawade (2015) in his study tried to explain the concept of MAKE IN INDIA which is progressing rapidly to provide enhanced economic growth. This study points out the importance of GM in overcoming challenges related to environmental protection and competitive business environment within Indian MSMEs. As an emerging concept, green manufacturing's several important areas of potential use and various factors for implementation are also highlighted. The current chapter makes the argument for accelerating the environmentally friendly strategy in the MSME sector in order to improve the environmental resilience of Indian products.

Baines et al. (2012) review the literature related to green production and its significance in improving environmental performance of manufacturing industries. The study summarizes the literature which covers various aspects of GM and its implementation in manufacturing industries. There are a number of factors that affect the implementation of GM in an industry. The results of the study indicate that top management commitment, training to employees, planning, and technology infrastructure are the key factors that positively affect the successful implementation of GM.

According to Singh et al. (2018), growing environmental awareness has forced businesses to innovate and lessen its environmental impact. The most recent sustainable manufacturing method, known as "green manufacturing" (GM), has the potential to address the majority of the problems now affecting the Indian manufacturing sector. GM makes the best use of the resources at hand and reduces waste in order to minimize its impact on the environment. Small and medium-sized businesses (SMEs) may have been excluded from the regulatory and social pressures for a variety of reasons. Compared to conventional manufacturing, green manufacturing is unique. It emphasizes the effects on the environment, government environmental policies, national and international environmental laws, stakeholder environmental action, and market forces. Environmental restrictions or standards that are appropriate can spur green technologies that actually save costs, boost productivity, or raise companies' competitiveness.

According to Zhou et al. (2008), a green product has six features: it is developed and designed, manufactured, packaged, sold and transported, used and maintained, and recycled. Additionally, a system, technique, and method for evaluation were built in order to assess the project for designing green products. About 37 indicators and 6 attributes—technical, environmental, resource, energy, economic, and social—were included in the evaluation index system. These regulations and opinions become the requirements of the operational capability plan. Materials, production processes, packaging and transportation, usage, waste, and recovery are among a few of the characteristics that Taghaboni-Dutta et al. (2010) suggested as making up green products. The designer and stakeholder perspectives were merged by Madu et al. (2002) to create a hierarchical framework for environmentally conscious design. The approach assessed environmental impact in addition to product attributes. When adopting eco-designs that affect the environmental performance of the company, Zailani et al. (2012) attempted to analyse the internal proactive environmental strategy and external institutional forces.

In their study, Zhang et al. (2005) addressed how R&D alliances impact various businesses. They are impacted by a variety of things. Participating in R&D alliances is advantageous for small and medium-sized businesses' green manufacturing. Capital investment has a good effect on green manufacturing; as more money is invested in these activities, the advantages come about more quickly. Inputs from human resources in green manufacturing are beneficial. It implies that research findings are

TABLE 8.1
Barriers in Implementing Green Manufacturing

S. No	Barriers
1	Employee demands
2	Lack of health and safety facilities
3	Inappropriate company culture
4	Lack of innovation
5	Lack of financial profit
6	Inappropriate stakeholder relations
7	Lack of environmental concerns
8	Lack of competitive advantage
9	Negative market trend
10	Lack of internal motivations
11	Environmental protection laws and regulation
12	Lack of green image
13	Lack of competitors green strategy
14	Lack of management commitment and participation
15	Lack of employee empowerment and involvement
16	Inappropriate reduce, reuse, remanufacturing methods
17	Inappropriate disposal of hazardous materials
18	Lack of green product
19	Current legislation
20	Lack of awareness
21	Lack of understanding of green productivity
22	Lack of regulation-driven motivation
23	Financial constraints
24	Technological risks
25	Lack of training facilities

produced more quickly the more technical people participate in green manufacturing activities. Green manufacturing is positively impacted by technological intensity. Companies with high levels of green manufacturing practices perform better than those with low levels of green manufacturing practices. Alliances in R&D for green manufacturing are beneficial. The firms that participate in R&D alliances outperform those that conduct independent research and development in terms of green manufacturing. The difficulties in implementing green manufacturing are outlined in Table 8.1 by a number of authors in earlier literature.

In addition, Table 8.2 categorizes the identified barriers in the implementation of GM into five major groups that have been highlighted by various authors.

8.3 DESIGN OF STUDY/RESEARCH METHODOLOGY

Work on this study has been done at MSMEs in Northern India. It attempts to determine and assess the main obstacles to the adoption of GM in MSMEs. The questionnaire

TABLE 8.2
Five Major Categories of Barriers

S. No	Categories of Barriers	Barriers
1	Lack of organizational encouragement	Employee demands
		Lack of health and safety facilities
		Inappropriate company culture
		Lack of innovation
		Lack of financial profit
2	Lack of awareness about potential benefits	Inappropriate stakeholder relations
		Lack of environmental concerns
		Lack of competitive advantage
		Negative market trend
		Lack of internal motivations
3	Lack of top management commitment	Environmental protection laws and regulation
		Lack of green image
		Lack of competitors green strategy
		Lack of management commitment and participation
		Lack of employee empowerment and involvement
4	Lack of waste management	Inappropriate reduce, reuse, remanufacturing methods
		Inappropriate disposal of hazardous materials
		Lack of green product
		Current legislation
		Lack of awareness
5	Lack of facilities	Lack of understanding of green productivity
		Lack of regulation driven motivation
		Financial constraints
		Technological risks
		Lack of training facilities

was created with responses based on the five-point Likert Scale following a thorough literature research and expert advice. Peer evaluation of the questionnaire by academics, business managers, and industrialists/entrepreneurs served to validate it. Three stages of data gathering were conducted in order. Various MSMEs were randomly chosen from directories of MSMEs in Northern India and contacted by phone, in-person interview, or email to inform them of the survey work being done in the area. In order to get the best results possible in the first round, the owners or employees of the organizations were also requested for their permission to participate in the survey. The questionnaire and the letter of intent were sent to them in the subsequent/second step by email or postal delivery. Then, in the final round, phone calls and reminders were made to them to complete the information and mail it back. Personal interviews with business owners or top executives produced the best results out of all the techniques used. A total of 84 replies in all were gathered over the course of 18 months (September 2022 to February 2023). The ultimate response rate was not greatly impacted by a few tardy responses. A few missing values were detected in 11 replies, making them only partially complete. About 72 replies remained for examination after the rejection of unacceptable ones, which was adequate for data analysis.

The questionnaire included 25 enquiries about GM obstacles to execution, which were broken down into 5 main categories and graded using a Likert scale. According to the replies received, the mean score illustrates the weighting given to each particular barrier in the implementation of GM. In the case of India, the government has created rules and policies to encourage the adoption of GM in MSMEs and demonstrate its effects. However, in order to put them into practise, industrialists still confront numerous barriers. The list of barriers includes a wide variety of obstacles that organizations may encounter. With the use of Pareto analysis and the AHP method, their actual impact as a barrier to the deployment of GM has been assessed.

8.4 ASSESSING AND DISCUSSION OF RESULTS USING AHP APPROACH

The findings of a thorough survey carried out in the industrial sector are presented in this section. The survey's goal is to evaluate the severity of the issues encountered by particular industries. The survey investigates the current state of industry implementation challenges for GM. The survey's research focuses on the issues facing each industry sector based on their significance and research-backed effects on manufacturing output.

8.4.1 Pareto Analysis

By equating the total frequency to 100%, the Pareto analysis ranks the data classification in descending order of frequency of occurrence from highest to lowest. Below is a presentation of the barriers to Pareto analysis from the chosen publications. Frequency and cumulative percentage of barriers are calculated and given in Table 8.3.

Figure 8.1 shows the frequency and cumulative percentage of barriers which when added together gives a sum total of hundred.

Table 8.4 represents a sorted list of vital few which consists of lack of organizational encouragement, lack of awareness about potential benefits, lack of top

TABLE 8.3
Frequency and Cumulative Percentage of Barriers

Category	Frequency	Cumulative Frequency	Cumulative Percentage
Lack of organizational encouragement	24	24	22.02
Lack of awareness about potential benefits	22	46	42.2
Lack of top management commitment	23	69	63.30
Lack of waste management	21	90	82.57
Lack of facilities	19	109	100.00

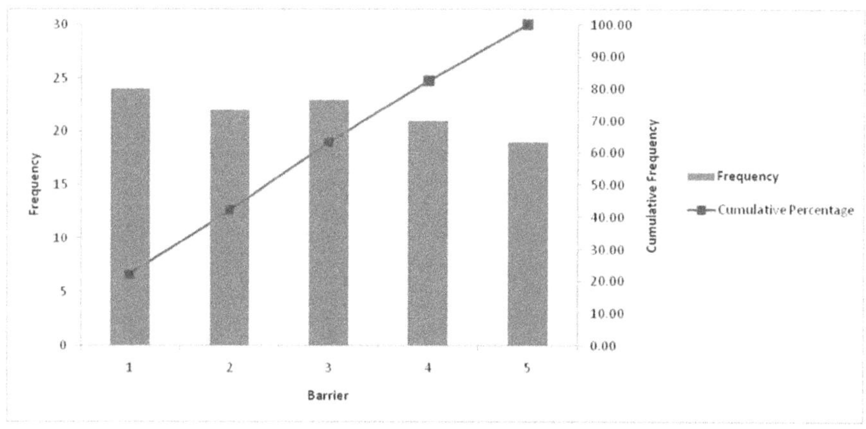

FIGURE 8.1 Pareto chart of barriers.

TABLE 8.4
Sorted List of Vital Few and Trivial Many Barriers

S.No	Vital Few	S.No	Trivial Many
1	Lack of organizational encouragement	1	Lack of waste management
2	Lack of awareness about potential benefits	2	Lack of facilities
3	Lack of top management commitment		

management commitment, and trivial many which consist of lack of waste management and lack of facilities.

8.4.2 Analytic Hierarchy Process (AHP) approach

Finding the optimum solution to complicated decision-making problems has been done via the Analytical Hierarchy Process (AHP). It has been utilized in a variety of industries, including banking, marketing, sports, education, healthcare, and public policy (Saaty, 1990, 1994a,b). Because the key concern or criteria weights generated by the AHP approach are based on human judgements and not artificial scales, compared to other approaches, AHP gives better and improved results (Golden et al., 1989). The other parts can be compared to one another after the hierarchy has been created. The evaluation is transformed into numerical numbers, making the entire issue simple to address. Singh et al. (2019) also looked at how Indian MSMEs were implementing technological innovation.

After receiving scores from specialists from various organizations, the key influential barriers were selected before moving on to the AHP phases. There were variances in the respondents' scores due to their situations, work cultures, experiences, products, and so on, but these barriers were selected after a thorough literature analysis. In order to use AHP, a problem must be well characterized, broken down into its major

TABLE 8.5
Brief Profile of the Experts for Deploying AHP Approach

S. No.	Position in Organization	Sector/Product	Experience (years)
1	Deputy Director	Manufacturing	14
2	HR Head	Electrical goods manufacturing	12
3	Senior Engineer	Automotive parts manufacturing	18
4	Professor and GM Expert	Academia	21
5	Proprietor	Automotive parts manufacturing	10
6	Senior Engineer	Hand tools manufacturing	12
7	Owner	Manufacturing	11
8	Manager	CAD/CAM manufacturing	18
9	Proprietor	CNC machining	16
10	Manager	Rubber parts manufacturing	8
11	Professor and GM Expert	Academia	14
12	Owner	CAD/CAM manufacturing	16

components, and then further divided into pairs. Before specialists from academia and industry, the 25 most significant barriers were presented in order to be evaluated. The brief profile of the experts is shown in Table 8.5.

The comparisons were made and the experts were requested to fill these on the basis of scale for AHP judgement as mentioned in the below list:

- If i and j are equally essential, the rating is 1.
- If *i* is marginally more significant than *j*, rate it as 3.
- Rate as 5 if *i* is more crucial than *j*.
- If *i* is significantly more important than *j*, the rating is 7.
- If *i* is unquestionably more significant than *j*, the rating is 9.
- Rate 2, 4, 6, or 8, which are neutral numbers in the range of evaluations.

8.4.3 AHP Hierarchy

The hierarchy for any decision-making in AHP is produced by breaking down the present statement into many hierarchies (levels) of choice elements. In the current work, two alternatives success and failure are considered for green manufacturing barriers. These are:- EPLR, GI, CGS, MCP, EEI, RRR, DHM, GP, CL, AW, UGP, RDM, FC, TR, LTF. Additionally, this procedure can produce priority weights (scores) for comparing each choice.

8.4.4 Pairwise Comparison

The relative relevance of the traits in relation to one another is calculated in this comparison. The reciprocal judgement matrix that results from the comparison of those traits is n × n. A 15 × 15 matrix was created for this, depending on the amount of attributes, as illustrated. Table 8.6 shows pairwise comparison and the sum total for the different barriers.

TABLE 8.6
Pairwise Comparison

Name	EPLR	GI	CGR	MCP	EEI	RRR	DHM	GP	CL	AW	UGP	RDM	FC	TR	LTF
EPLR	1	4	2	1	3	4	4	3	2	1	2	2	3	4	3
GI	0.25	1	3	4	1	2	2	3	1	1	4	2	4	3	3
CGS	0.5	0.33	1	2	1	2	3	1	4	4	2	3	2	2	2
MCP	1	0.25	0.5	1	2	2	4	1	3	3	2	2	1	1	4
EEI	0.33	1	1	0.5	1	3	3	1	2	2	2	3	2	4	2
RRR	0.25	0.5	0.5	0.5	0.33	1	2	2	3	3	4	1	2	3	2
DHM	0.25	0.5	0.33	0.25	0.33	0.5	1	4	3	1	2	3	3	3	3
GP	0.33	0.33	1	1	1	0.5	0.25	1	2	1	2	2	2	3	2
CL	0.5	1	0.25	0.33	0.5	0.33	0.33	0.5	1	4	2	2	4	4	2
AW	1	1	0.25	0.33	0.5	0.33	1	1	0.25	1	3	4	1	2	2
UGP	0.5	0.25	0.5	0.5	0.5	0.25	0.5	0.5	0.5	0.33	1	1	3	2	1
RDM	0.5	0.5	0.33	0.5	0.33	1	0.33	0.5	0.5	0.25	1	1	2	3	3
FC	0.33	0.25	0.5	1	0.5	0.5	0.33	0.5	0.25	1	0.33	0.5	1	4	4
TR	0.25	0.33	0.5	1	0.25	0.33	0.33	0.33	0.25	0.5	0.5	0.33	0.25	1	2
LTF	0.33	0.33	0.5	0.25	0.5	0.5	0.33	0.5	0.5	0.5	1	0.33	0.25	0.5	1
Total	7.33	11.58	12.16	14.16	12.75	18.25	22.41	19.83	23.25	23.58	28.83	27.16	30.5	39.5	36

The priority vector, a normalized eigenvector of the matrix, is computed after the comparison matrix has been created. The normalized matrix is created by dividing each column by the total number of entries in that column.

The normalized value Nij is determined as follows:

$$n_{ij} = a_{ij} / \Sigma\, a_{ij} \text{ where } i = 1 \text{ to } n$$

For each property, the approximation of the priority weight (w1, w2, w3, and wj) is obtained as

$$w_j = 1/n * \Sigma\, a_{ij} \text{ where } i = 1 \text{ to } n$$

The normalized matrix of various variables is shown in Table 8.7, along with priority weights determined from a pairwise comparison matrix.

8.4.5 Consistency Checks

The relative weights, which would also display the criteria's eigenvalues, should confirm

$$AWi = \lambda\, max\, Wi, \text{ i} = 1,2, 3, \ldots, \text{n}$$

where max delivers the largest eigen value and A represents the pairwise comparison decision matrix.

The pairwise comparison discrepancies are measured by the consistency index (CI).

$$\text{CI} = \frac{\lambda\, max - \text{n}}{\text{n} - 1}$$

Calculations for the consistency ratio (CR) include

$$\text{CR} = \text{CI/RI}$$

To be considered acceptable, CR must be close to 0.10 (10%).

Table 8.8 shows the random consistency index (RI) which ranges between minimum value of 0.52 to maximum value of 1.58.

Table 8.9 represents the consistency index having value of 0.195.

Table 8.10 shows the value of consistency measure for the various barriers in implementing green manufacturing.

8.4.6 Alternatives' Priority Weights in Relation to an Attribute

Priority weights are used to gauge how much an attribute is valued over the alternative (success or failure). An organization is more likely to succeed if one of its attributes is present and strong than if the other is there but weak. Priority weights involve multiplying each alternative's weight evaluation by the attribute weight vector in the

TABLE 8.7
Normalized Matrix

Factor	EPLR	GI	CGS	MCP	EEI	RRR	DHM	GP	CL	AW	UGP	RDM	FC	TR	LTF	Weight
EPLR	0.136	0.345	0.164	0.071	0.235	0.219	0.178	0.151	0.086	0.042	0.069	0.074	0.098	0.101	0.083	0.137
GI	0.034	0.086	0.247	0.282	0.078	0.110	0.089	0.151	0.043	0.042	0.139	0.074	0.131	0.076	0.083	0.111
CGS	0.068	0.029	0.082	0.141	0.078	0.110	0.134	0.050	0.172	0.170	0.069	0.110	0.066	0.051	0.056	0.092
MCP	0.136	0.022	0.041	0.071	0.157	0.110	0.178	0.050	0.129	0.127	0.069	0.074	0.033	0.025	0.111	0.089
EEI	0.045	0.086	0.082	0.035	0.078	0.164	0.134	0.050	0.086	0.085	0.069	0.110	0.066	0.101	0.056	0.083
RRR	0.034	0.043	0.041	0.035	0.026	0.055	0.089	0.101	0.129	0.127	0.139	0.037	0.066	0.076	0.056	0.070
DHM	0.034	0.043	0.027	0.018	0.026	0.027	0.045	0.202	0.129	0.042	0.069	0.110	0.098	0.076	0.083	0.069
GP	0.045	0.029	0.082	0.071	0.078	0.027	0.011	0.050	0.086	0.042	0.069	0.074	0.066	0.076	0.056	0.058
CL	0.068	0.086	0.021	0.024	0.039	0.018	0.015	0.025	0.043	0.170	0.069	0.074	0.131	0.101	0.056	0.063
AW	0.136	0.086	0.021	0.024	0.039	0.018	0.045	0.050	0.011	0.042	0.104	0.147	0.033	0.051	0.056	0.058
UGP	0.068	0.022	0.041	0.035	0.039	0.014	0.022	0.025	0.022	0.014	0.035	0.037	0.098	0.051	0.028	0.037
RDM	0.068	0.043	0.027	0.035	0.026	0.055	0.015	0.025	0.022	0.011	0.035	0.037	0.066	0.076	0.083	0.042
FC	0.045	0.022	0.041	0.071	0.039	0.027	0.015	0.025	0.011	0.042	0.012	0.018	0.033	0.101	0.111	0.041
TR	0.034	0.029	0.041	0.071	0.020	0.018	0.015	0.017	0.011	0.021	0.017	0.012	0.008	0.025	0.056	0.026
LTF	0.045	0.029	0.041	0.018	0.039	0.027	0.015	0.025	0.022	0.021	0.035	0.012	0.008	0.013	0.028	0.025

TABLE 8.8
Random Consistency Index

n	3	4	5	6	7	8	9	10	11	12	13	14	15
RI	0.52	0.89	1.11	1.25	1.35	1.40	1.45	1.49	1.51	1.54	1.56	1.57	1.58

TABLE 8.9
Consistency Index

6	CI	RI	CR
17.73	0.195	1.58	0.123

TABLE 8.10
Consistency Measure

Factor	Value	Rank
EPLR	18.072	**5**
GI	17.828	**7**
CGS	18.691	**1**
MCP	18.273	**4**
EEI	18.336	**3**
RRR	18.381	**2**
DHM	17.827	**8**
GP	17.515	**9**
CL	17.838	**6**
AW	17.135	**13**
UGP	17.399	**11**
RDM	17.062	**14**
FC	16.866	**15**
TR	17.337	**12**
LTF	17.413	**10**

evaluation rating matrix and adding the results across all attributes. Success in pairs depends on how much one is superior to the other. According to how much better one is than the other at satisfying each criterion, a pair is successful.

Decision index of success = (0.667×0.137) + (0.667×0.111) + (0.752×0.092) + (0.667×0.089)+ (0.800×0.083) + (0.752×0.070) + (0.667×0.069) + (0.752×0.058) + (0.752×0.063) +(0.800×0.058) + (0.667×0.037) + (0.752×0.042) + (0.800×0.041) + (0.800×0.026) +(0.667×0.025)

=0.7217 or 72.17%

TABLE 8.11
Priority Weights with Respect to Success and Failure

Sr No		Success	Failure	Weight
1	**Success**	1	2	0.667
	Failure	0.5	1	0.333
2	**Success**	1	2	0.667
	Failure	0.5	1	0.333
3	**Success**	1	3	0.752
	Failure	0.33	1	0.248
4	**Success**	1	2	0.667
	Failure	0.5	1	0.333
5	**Success**	1	4	0.800
	Failure	0.25	1	0.200
6	**Success**	1	3	0.752
	Failure	0.33	1	0.248
7	**Success**	1	2	0.667
	Failure	0.5	1	0.333
8	**Success**	1	3	0.752
	Failure	0.33	1	0.248
9	**Success**	1	3	0.752
	Failure	0.33	1	0.248
10	**Success**	1	4	0.800
	Failure	0.25	1	0.200
11	**Success**	1	2	0.667
	Failure	0.5	1	0.333
12	**Success**	1	3	0.752
	Failure	0.33	1	0.248
13	**Success**	1	4	0.800
	Failure	0.25	1	0.200
14	**Success**	1	4	0.800
	Failure	0.25	1	0.200
15	**Success**	1	2	0.667
	Failure	0.5	1	0.333

As a result, the failure index is = 1−0.7217 = 0.278 or 27.8%.

Table 8.11 shows the priority weights with respect to success and failure of the formulated model.

This shows that applying the practice while taking into account barriers has a success rate of 72.17% and a failure rate of 27.88%. Practitioners and project managers can take these findings into consideration for the successful implementation of green manufacturing.

8.5 CONCLUSIONS

It is apparent that GM has several advantages when being implemented in MSMEs. It can raise MSMEs' competitive and environment-adaptive levels. In developing countries like India, adoption and implementation of GM are urgently needed. MSMEs have numerous barriers when attempting to use GM. This research uses the AHP approach to analyse these barriers. The most significant and impactful hurdles require more attention in order to move forward with the deployment of GM, even though all barriers are significant and should be addressed with a solution.

Lack of competitors' green strategy, inappropriate reduce, reuse, and remanufacturing methods, lack of employee empowerment and involvement, and lack of management commitment and participation are the major barriers which obstruct the application of GM in MSMEs. The managers or industrialists of various nations who are on the same route or will follow shortly can benefit from this study in a variety of ways. The major barriers were examined in-depth analysis. There has also been a model for successfully deploying GM in MSMEs. This study's findings indicate that when using green practises while taking into account barriers, there is a 72.17% success rate and a 27.88% failure rate. These results can be taken into account by practitioners and project managers to successfully implement green manufacturing.

Since the present study considers a larger variety of criteria in order to properly implement the GM and remove the main barriers, the proposed model looks to be an improvement over the models that have previously been investigated in the literature.

Future scope of work: This enquiry was based on a survey, which was then evaluated by academic and professional specialists, but alternative methodologies, such as PROMETHEE, WPM, VIKOR, and so on, could be used for the investigation. Additionally, the study can be conducted in various areas and regions of the nation.

REFERENCES

Agan, Y., Acar, M., and Borodin, A. (2013). "Drivers of environmental processes and their impact on performance; A study of Turkish SMEs", *Journal of Cleaner Production*, 51, 23–33.

Baines, T., Brown, S., Benedettini, O., and Ball, P. (2012). "Examining green production and its role within the competitive strategy of manufacturers", *Journal of Industrial Engineering and Management*, 5(1), 53–87.

Castka, P., Gabzdylova, B., and Raffensperger, J. (2009). "Sustainability in the New Zealand wine industry: drivers, stakeholders and practices", *Journal of Cleaner Production*, 17(11), 992–998.

Daily, B., Bishop, J., and Massoud, J. (2012). "The role oftraining and empowerment in environmental performance: a study of Mexican maquiladora industry", *International Journal of Operations & Production Management*, 32(5), 631–647.

Ferreira Inês, A., et al. (2023). "Assessing the impact of fusion-based additive manufacturing technologies on green supply chain management performance", *Journal of Manufacturing Technology Management*, 34(1), 187–211.

Ghazilla, R., Sakundarini, N., Abdul-Rashid, S., Ayub, N., Olugu, E., and Musa, S. (2015). "Drivers and barriers analysis for green manufacturing practices in Malaysian SMEs: a preliminary findings", *Procedia CIRP*, 26, 658–663.

Golden, B. L., Wasil, E. A., & Harker, P. T. (1989). The analytic hierarchy process. Applications and Studies, Berlin, Heidelberg, 2(1), 1–273.

Hicks, C. and Dietmar, R. (2007), "Improving cleaner production through the application of environmental tools in China", *Journal of Cleaner Production*, 15(5), 395–408.

Hutchinson, A. and Chaston, I. (1994), "Environmental management in devon and cornwall's small and medium sized enterprise sector", *Business Strategy and the Environment*, 3(1), 15–22.

Kothawade, N.S. (2015), "Green manufacturing: solution for Indian climate change commitment and make in India aspirations", *International Journal of Science and Research,* 2319–7064.

Madu, C.N., Kuei, C., and Madu, I.E. (2002), "A hierarchic metric approach for integration of green issues in manufacturing: a paper recycling application", *Journal of Environmental Management*, 64(3), 261–272.

Mohanty, R. P., & Prakash, A. (2014). Green supply chain management practices in India: a confirmatory empirical study. Production & Manufacturing Research, 2(1), 438–456.

Saaty, T.L. (1990), "How to make a decision: the analytical hierarchy process", *European Journal of Operations Research*, 48(1), 9–26.

Saaty, T.L. (1994a), "How to make a decision: the analytic hierarchy process", Interfaces, Vol. 24 No. 6, pp. 19–43.

Saaty, T.L. (1994b), "Highlights and critical points in the theory and application of the analytic hierarchy process", *European Journal of Operational Research*, 74(3), 426–447.

Singh, A., Jha, S., and Prakash, A. (2014), "Adoption of green manufacturing (GM) in selected Indian industries", *International Journal of Applied Engineering Research*, 9, 20383–20404.

Singh, C., Singh, D., & Khamba, J. S. (2021). Analyzing barriers of Green Lean practices in manufacturing industries by DEMATEL approach. *Journal of Manufacturing Technology Management*, 32(1), 176–198.

Singh, D., Khamba, J.S., and Nanda, T. (2018), Problems and prospects of Indian MSMEs: a literature review. *International Journal of Business Excellence*, 15(2), 129–188.

Singh, D., Khamba, J.S., and Nanda, T. (2019), Justification of technology innovation implementation in Indian MSMEs using AHP. *International Journal of Services and Operations Management*, 32(4),522–538.

Singh, M.D., and Thakar, G. D.(2018), Green manufacturing practices in SMES of India, *A Literature Review, Industrial Engineering Journal*, XI(3), 37–45.

Taghaboni-Dutta, F., Trappey, A.J.C. and Trappey, C.V. (2010), An XML based supply chain integration hub for green product lifecycle management, *Expert Systems with Applications*, 37(11), 7319–7328.

Xu, J., (2017). Research on Green Manufacturing Innovation Based on Resource Environment Protection Sichuan Academy of Social Sciences, Chengdu, IOP Publishing IOP Conf. Series: Earth and Environmental Science 94.

Young, P., Byrne, G. and Cotterell, M. (1997), Manufacturing and the environment, *The International Journal of Advanced Manufacturing Technology*, 13(7), 488–493.

Zailani, S., Eltyeb, T., Hsu, C., and Tan, K. (2012), The impact of external institutional drivers and internal strategy on environmental performance, *International Journal of Operations & Production Management*, 32(6), 721–745.

Zhang, X., Aguilar, E., Sensoy, S., Melkonyan, H., Tagiyeva, U., Ahmed, N., ... & Wallis, T. (2005). Trends in Middle East climate extreme indices from 1950 to 2003. Journal of Geophysical Research: Atmospheres, 110(D22).

Zhou, X., Zhang, Q.S., Zhang, M., and Li, X. (2008), Research on evaluation and development of green product design project in manufacturing industry, International Conference on Wireless Communications, Networking and Mobile Computing, Crete, Greece, 6–8 August 2008, pp. 1–5

Zhu, Jianhua, et al. (2023), The choice of green manufacturing modes under carbon tax and carbon quota. *Journal of Cleaner Production* , 384, 135336.

9 Green Manufacturing Implementation in Industries
Case Studies

Pankaj Rana, Dharmpal Deepak, and Sulakshna Dwivedi

9.1 INTRODUCTION

Green manufacturing is a modern manufacturing process that assures the product's quality, features, and price while also taking into account the environment's impact and resource efficiency. Green manufacturing reduces environmental contamination across the whole product life cycle, from design to manufacture to use to waste, maximizing resource utilization and minimizing energy usage (Zheneng 2010; Yinbao 2014). In order to protect the environment from raw materials, production, scrap, and other elements, the conventional manufacturing model uses an open loop system that uses terminal treatment. Due to the manufacturing sector's explosive growth, environmental issues have gained attention and are now a top priority for all countries. The manufacturing industry is compelled by the "green wave" to renounce its traditional techniques of production and embrace green manufacturing technology (Lin and Hao 2020).

The process of making green products while utilizing green resources and energy is referred to as "green manufacturing." Green energy refers to the effective use of resources over the course of a product's entire life cycle. The three main issues facing the globe now are population, environment, and resource depletion. The production of necessary goods is growing daily to keep up with the demands of the current population. Such activities have a very negative influence on the environment, and the majority of the resources that are available to supply such demands are non-renewable, posing a serious threat to human survival and development. Waste generated during the process of transforming raw materials into finished commodities seriously contaminates the environment. Modern manufacturing practices known as "green manufacturing" take into account proper resource management and environmental impact (Figure 9.1). Additionally, green production is based on the principles of ensuring quality, reducing waste, and optimizing resource use (Lin and Hao 2020).

 DOI: 10.1201/9781003449225-9

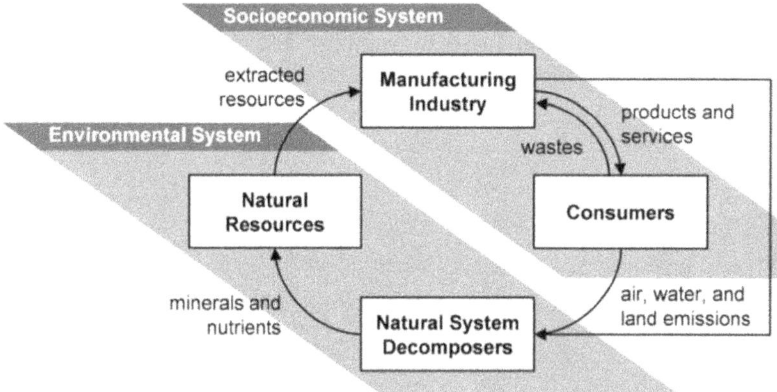

FIGURE 9.1 Contribution of manufacturing sectors to a sustainable system. (Figure reused after permission of the publisher).

Source: Paul et al. (2014).

The three primary issues facing the world today are all present in green manufacturing. The primary characteristics of green manufacturing are its methodical comparison of traditional and green methods, emphasis on prevention, adherence to financial regulations, and consideration of efficiency. The entire life cycle of the product is the primary factor in green production (Singh et al. 2022). With 40% of its installed electrical capacity coming from non-fossil fuel sources, India is the third-largest generator of renewable energy in the world (Ministry of New and Renewable Energy 2022)

In 2020, the demand for all other fuels decreased while the usage of renewable energy increased by 3% (Figure 9.2). The main factor was an increase of about 7% in the amount of electricity produced from renewable sources. Even with decreasing electricity demand, supply chain issues, and building delays in many regions of the world, long-term contracts, priority access to the grid, and continual installation of new plants supported the rise of renewable energy sources. As a result, from 27% in 2019 to 29% in 2020, the proportion of renewable energy sources in the world's electricity generation increased. While the use of bioenergy in industry increased by 3%, the fall in the use of biofuels—due to decreasing oil demand—largely countered the increase (Global Energy Review 2021 – Analysis – IEA). The highest year-over-year rise in renewable power generation since the 1970s is expected to occur in 2021 when it will increase by more than 8% to 8300 TWh. Two-thirds of the growth in renewables is anticipated to come from solar PV and wind (Global Energy Review 2021 – Analysis – IEA).

9.2 GREEN MANUFACTURING CHARACTERISTICS

a. The following are the significant characteristics of green manufacturing:
   The primary difference between green manufacturing and conventional manufacturing is that while green manufacturing ensures the basic functions

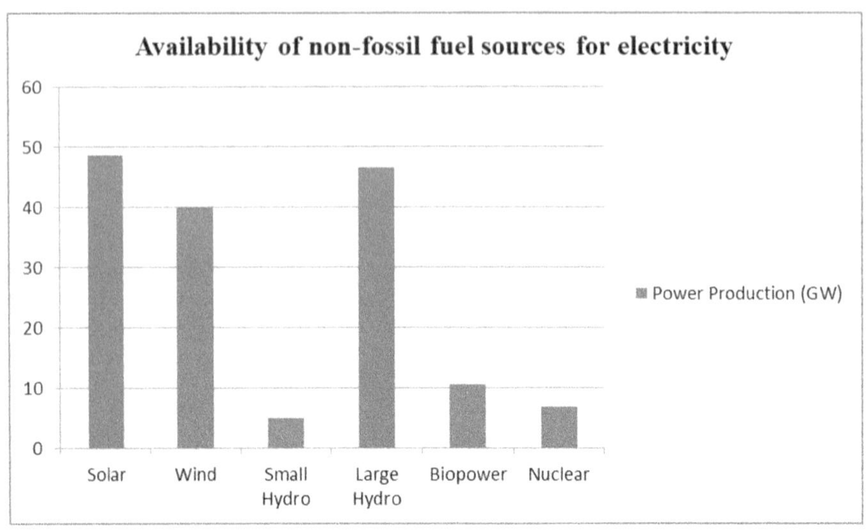

FIGURE 9.2 Renewable energy capacity installed in India.

Source: Renewable Energy in India (2022).

of the industrial system, it also ensures the least amount of environmental damage.

i. Make an effort to prevent:
During the product-production process, waste reduction or elimination is given top priority in green manufacturing.
ii. Be active:
Green manufacturing must incorporate product and process criteria in order to satisfy the demands of the anticipated development while protecting the ecological environment and ensuring resource efficiency.
iii. Economic commitment:
Green manufacturing techniques help to reduce energy and raw material consumption, as well as the costs of waste management and disposal, production expenses, product economy, and market competitiveness.
iv. Keep an eye on effectiveness:
Green manufacturing has moved away from the treatment of end-of-products in favor of continuous product and production process control, which has also heavily incorporated recycling technologies of renewable resources, energy, and materials to effectively prevent secondary pollution (Tianyi 2014; Xianguang 2014).

b. The green manufacturing system can be summed up in three ways, two goals, and three contents:

Three ways of green manufacturing:

i. Legal Training Method
ii. Managing Style
iii. Technical Method

Two goals of green manufacturing:

i. Comprehensive use of available resources
ii. Protecting the environment

Three contents of green manufacturing:

i. Green manufacturing
ii. Green manufacturing technique
iii. Sustainable goods

9.3 TOOLS OF GREEN MANUFACTURING

By producing products that are kind to the environment and profitable in a sustainable manner, green manufacturing supports the client. These environmentally friendly procedures cover a wide range of methods for reducing waste, such as reusing and recycling, reengineering, remanufacturing, replacing potentially hazardous materials, and internally consuming trash. Green manufacturing, in general, focuses on reducing negative environmental effects by creating and supplying such items from precise production, use, and disposal. The implementation of such practices in the market in accordance with client needs is specified by strategic practices. There are several stages to green manufacturing, including supply chain management, process planning, and green technologies (Logesh and Balaji 2021).

Typically, these are top management-established environmental awareness plans, goals, and policies. Generally, it takes the participation, involvement, knowledge of technology, and support of the stack holders to build a strategy plan (Ahn 2016). The 3R (Reuse, Reduce, and Recycle) principle and other approaches, such as material processing, distribution, end use, material extraction, and tracking raw materials, have all been developed by researchers to use minimal resources, control pollution and emissions, track raw materials, waste segregation and reuse, change in product and process, and reduce the environmental impact of manufactured goods over their lifetime.

9.4 A DESCRIPTION OF THE INDIAN MANUFACTURING SECTOR

Indian manufacturing dates back to the nineteenth century, but after independence, in the second five-year plan, it started to rapidly modernize and flourish. The economic reforms of 1991, which eliminated import restrictions, increased foreign competition, led to the privatization of some governmental industries, liberalized the foreign direct investments (FDI) regime, improved infrastructure, and increased production of fast-moving consumer goods, had a significant impact on the Indian industrial sector. India, one of the world's fastest-growing economies, had the highest growth rates in the middle of the 2000s (Digalwar et al. 2013; Sangwan 2006). India has a 75% working population, according to a study report on the industrial sector. India has a large labor force and offers the necessary number of employment opportunities.

India has the ninth-largest economy in the world. Over 17% of India's workforce is employed in the industrial sector, which contributes about 27.6% of the country's GDP. India ranks 12th in the world in terms of nominal industry output in absolute terms. India ranks 11th in terms of imports and 17th in terms of exports worldwide. For the fiscal year 2011–2012, economic growth rates were between 7.5% and 8% (Digalwar et al. 2017). According to the Confederation of Indian Industry (CII), the Indian economy grew by 7.4% in 2009–2010, and industrial growth was predicted to reach 8.5% for 2010–2011. Manufacturing contributed 7.6% to the industrial sector's growth rate of 9.8% in 2006–2007. Manufacturing provided 7.4% of the industrial growth in 2005–2006, which was 9.0%. Manufacturing provided 6.6% of the industrial growth in 2004–2005, which was roughly 8.1%. About 7.1% of India's GDP in 2003–2004 came from industry. The manufacturing sector's contribution in 2002–2003 was about 5.7%. This demonstrates that the GDP growth rate of the Indian economy has been increasing in recent years (Toke and Kalpande 2019).

In fact, India is advancing toward becoming an industrialized economy. In recent years, India's economy has continued to be significantly influenced by the manufacturing sector. However, the rapid industrialization had a detrimental effect on the environment due to an increase in pollution, waste, and the quick depletion of natural resources (Bhanot et al. 2017). It is abundantly obvious that the manufacturing sector is the main cause of the environmental issue. Environmental concerns have grown significantly in importance for the public and the government (Thanki et al. 2016). In order to implement sustainability in the industrial sector, it is necessary to identify the underlying variables that can offer an appropriate framework. In order to address environmental concerns in the manufacturing sector, the study must look into the extent of green manufacturing adoption and implementation in Indian manufacturing sectors.

India was placed sixth as of 2018 with a contribution to the global GDP of 16% and industrial output of 3%. This is a far cry from some of the smaller nations like South Korea, Taiwan, and Vietnam, who contribute substantially more to their own GDP while also having a similar percentage of global manufacturing. In an effort to move from an agrarian to an industry-centric economy, India somehow ended up with a service sector-driven economy, which today (as of 2021) contributes 49.9% to the GDP. This is the main cause of poor manufacturing production (Pulicherla et al. 2022).

9.5 SIGNIFICANCE OF GREEN MANUFACTURING IN THE INDIAN AND GLOBAL CONTEXT

Efforts must be made to protect the environment, and preventive measures will not only address managerial, social, and economic challenges but also manufacturing, according to the 2015 United Nations Climate Change Conference and its declarations. Everyone anticipates that economies with major environmental weight on the world stage, such as China and India, will participate in appropriate and rigorous green manufacturing-based frameworks and other ways. Because of this, the study has global significance. This study's main impetus is India's quickly changing manufacturing environment. There are now five discernible developments happening in India (Seth and Panigrahi 2015). The availability of young, less expensive skilled labor, business globalization, the influx of cheap goods from developing countries (such as China and Korea) boosting competition, new e-business models influencing the manufacturing and business situation, and government policies that support FDIs with a focus on "make in India" are a few of these. These factors compel business executives to use India as a center for manufacturing and sourcing rather than just for the purpose of advertising their products (Mohanty et al. 2007; Rehman et al. 2013).

These developments have a swirling impact on the markets and the manufacturing industry, increasing green and quality consciousness. As a result of these modifications, academics and practitioners are more likely to experiment with cutting-edge techno-managerial strategies, like green manufacturing, both separately and in combination. They are also more likely to comprehend the relationships between critical success factors (CSFs) and performance indicators. This transformation is both creating disaster and offering a huge opportunity to test out new techno-managerial models, methodologies, and frameworks because of green-based advancements, sustainability needs, and rigorous restrictions (Seth and Gupta 2005).

Wastes through value stream mapping to deliver value are anticipated to take center stage along with pressures for sustainability that are green in nature, rising quality standards in the Indian setting, and the adoption and implementation of modern frameworks that target quality issues.

9.6 RELEVANT TECHNOLOGIES OF GREEN MANUFACTURING

a. **Modern design technology:** Modern design technology will be the primary related technology of green manufacturing since green design is the foundation of green manufacturing.
b. **Modern manufacturing equipment and technology:** In order to realize high quality and efficiency, low consumption, and clean production, as well as to ensure product quality and market competition, modern manufacturing equipment and technology are the foundations of the machinery and processes of the green manufacturing system. As such, it is a crucial pillar of green manufacturing.
c. **Environmental engineering technology:** Green manufacturing aims to use resources as efficiently as possible and generate the least amount of waste possible in order to substantially reduce its harmful effects on the environment. Environmental engineering technology will therefore play a significant supporting role in the area of green manufacturing.
d. **Environmental technology norms, laws, and guidelines for resource use:** Green manufacturing rules, resource utilization strategies, and environmental technology standards have grown in importance as a result of the growth of networked, integrated, and intelligent production. The government can create

economic policies and use the workings of the market economy to direct the implementation of green manufacturing. Green manufacturing involves the government's behavior in terms of legislative and administrative measures.

e. **System engineering technology:** Green manufacturing implementation is a challenging system engineering issue. The system engineering approach is used to measure and assess the production system's utilization of resources, pollution to the environment, and level of adoption of green manufacturing.
f. **Technology for supporting communication networks and databases:** The development of the communication network must be the foundation for establishing, managing, and maintaining the appropriate green design database and knowledge base in order to satisfy the demands of green design and manufacturing.
g. **Concurrent engineering technology:** Green design demands the same design goals, such as reducing utilization of resources, making products simple to disassemble and recycle, protecting the environment, and ensuring that products meet specifications regarding quality, performance, life, and cost from the start of the product design. This is where concurrent engineering plays a special role.

9.7 TRANSFORMATION TO GREEN MANUFACTURING

The term "green transformation" describes procedures used by businesses and sectors that have a smaller negative impact on the environment. Industrial transformation has been one of the International Human Dimensions Program's (IHDP) main science programs since the 1990s. Vellinga and Herb (1999) merged the IHDP into their new "Future Earth" effort. Manufacturing businesses can address these issues by concentrating on the three areas listed below:

a. **Green energy:** Green energy comprises the creation and consumption of cleaner energy, as well as the utilization of renewable energy sources including CNG, solar, wind, and biomass power, and increased operational efficiency with regard to energy.
b. **Green products:** The following phase in this shift is the creation of greener products. The buzzwords "recycled," "low carbon footprint," "organic," and "natural" are increasingly used to describe environmentally friendly items.
c. **Green processes in business operations:** Employing green business practices is the third area. This calls for the efficient utilization of vital resources, the reduction of waste production through lean operations, the reduction of carbon footprint, and water conservation. Green business practices increase productivity and cut costs.

Many businesses have begun integrating green efforts into their everyday activities. The following factors motivate these initiatives:

i. Rising energy and input costs.

ii. Consumer demand for green products is rising.
iii. As governments implement new, stronger environmental and waste management regulations, regulatory demands are growing.
iv. Technological developments create exciting opportunities for new companies.

From being viewed as an "essential evil" to a "good enterprise," green has changed people's perceptions. Companies that implement green initiatives will benefit from increased brand recognition, political support, and regulatory compliance, improved capacity to recruit and retain talent, improved customer retention, and potential cost savings. However, because the economics of green manufacturing are still poorly understood, achieving these benefits calls for a long-term commitment and making trade-offs with immediate goals.

9.8 AN APPROACH FOR IMPLEMENTING GREEN MANUFACTURING

a. **Obstacles to adopting green:** The commercial case for Green is still strong even under challenging market conditions. There is a growing knowledge of the necessity of going green, as well as the necessity of addressing green in all three areas: green processes, green goods, and green energy. However, businesses confront difficulties on a number of fronts, with establishing leadership for such an effort being the greatest difficulty. Firms have to make evolution from the following:

    i. Switching from a more constrained, frequently isolated, and focused approach to green activities,
    ii. Achieving legal compliance, creating an eco-advantage, and
    iii. Evaluating initiatives as commercial possibilities as opposed to seeing them as centers of cost.

This demands a significant transformation that, in order to be successful, asks for a structure that addresses the three main obstacles to taking significant action:

i. Companies lack a thorough understanding of the factors and problems that are important to them and their sectors, as well as what a sustainable future means to them.
ii. Finding a strong business case for sustainability or even modeling one is challenging for organizations. Most people do not prioritize these activities, and since costs and technologies are still developing, it is frequently difficult to understand the economics.
iii. Even businesses that implement green initiatives carry out these tasks as a sideline to their main operations and without integrating them into their corporate strategies. Because of this, they are unable to reap all of the advantages and their execution is poor.

b. **Assessing the economy as well as making strategic decisions:** Like any significant transformative endeavor, implementing a greener lifestyle needs businesses to fully comprehend the costs, advantages, and range of green initiatives at their disposal. Companies must choose their green efforts depending on both financial and strategic evaluations of the options they identify once this fact basis has been built.

Estimating the long-term "value" produced by these initiatives is necessary for an economic assessment. If not, some of the long-term advantages of going green won't be included in a company's case. It should cover all value creation drivers, from quantitative ones like pricing power and cost savings to qualitative ones like staff recruitment and engagement.

Making an economic assessment is just a small portion of the narrative. After presenting a convincing argument, businesses must decide strategically how green they want to be and why. The selection of initiatives may fluctuate based on the market environment and prospects for strategy differentiation, in addition to the underlying economics of the options. Potentially, businesses have the option of being:

i. **Planet indifferent:** a place where few measures are taken.
ii. **Green citizen:** where they just implement a few isolated initiatives that are the bare minimum required by consumers or regulators.
iii. **Green innovators:** where they make an effort to stay ahead of the curve when it comes to sustainability and change the focus from managing risks to top-line development and important business opportunities.

The third option commits the organization to a thorough green strategy and to making the most of the initiatives, but the first two options do not allow the company to fully harness the potential of green and are only relevant with a short-term lens.

c. **Implementation framework:** The path to being green is one of extensive change. The preparation and execution of the initiatives must receive sufficient attention if they are to be successful. Early victories and triumphs are crucial for creating momentum. It necessitates a top management team that is totally dedicated, stringent periodic reviews, and ongoing internal and external communication.

All three fields of actions—green energy, green products, and green processes—can be implemented using a straightforward three-step methodology.

a. **Plan:** The business strategy, future planning of resources, and budgeting procedures all need to take green efforts into consideration. For instance, businesses must make a thorough plan to utilize more green energy, change their product line to greener options, and restructure their operations to adopt greener procedures. Green aims and measurements must be included in a sustainability charter that is based on both short- and long-term objectives.

Organizations should create green indices or scorecards that measure the impact of the green initiatives they have implemented, set precise goals for the indices, and monitor their progress toward those goals.

b. **Execute:** Green needs to be integrated across the value chain and made a part of the core business with a solid plan in place and targets that have been clearly established and tracked.

   i. **Green energy:** High energy-consuming manufacturing enterprises need to switch to cleaner energy sources and make plans to use them more effectively. Lowering the energy consumption of operations can be accomplished in large part by installing captive wind or solar power generation units and employing energy-saving techniques like installing LED lighting or making better use of daylight in building design.
   ii. **Green products:** Organizations should evaluate their goods according to (a) how Green resources and power are being utilized, (b) how Green the product is during its entire period of use, and (c) how Green the production procedure is in order to move toward a Green product portfolio. Companies can evaluate the Green value of their product offerings by analyzing these criteria. Companies should set goals for this measure within the planning process itself, and then monitor their success over time.
   iii. **Using green business practices:** Organizations must progressively revamp the operational procedures utilized throughout the value chain. This can entail switching to more environmentally friendly manufacturing methods, making adjustments to cut waste, increasing reuse and recycling assets, and encouraging every supplier, channels, consumers, and staff to do similar actions.

c) **Communicate:** For green efforts to completely realize their full potential advantages, a well-designed promotion strategy is just as crucial as thoughtful execution. Campaigns to educate customers about the company's green product offers and its energy and operational practices can result in higher sales.

9.9 LITERATURE REVIEW

SN	Year	Author(s)	Findings	References
1	2015	S. Fore and Charles Mbohawa	With a particular focus on the sector, the study looked into how the cement business affects the environment in a low-income, developing nation. A method emphasizing cleaner production was employed in a case study approach, with the main focus being on issues like particle pollution and toxic gases.	(Fore and Mbohawa 2015)

(*continued*)

SN	Year	Author(s)	Findings	References
2	2015	Rameshwar Dubey and Sadia Samar Ali	The authors looked at the adoption of green manufacturing practices by Indian companies and how this impacted the efficiency of the larger supply chain.	(Dubey and Ali 2015)
3	2015	Krishna Jasti Naga Vamsi, Aditya Sharma and Shashikantha Karinka	Creating eco-friendly items is primarily done to prevent the collapse of industrial growth. Using information from a similar study, an effort has been made to identify 80 different qualities and 11 pillars that serve as a framework for "green product development."	(Vamsi et al. 2015)
4	2015	Kuldip Singh Sangwan and Varinder Kumar Mittal	The authors did research on the frameworks and topics related to green development, including where they came from, what they mean, how broad they are, and what they have in common. Although many researchers have merged all of the models and methodologies, they discovered that standardization in the methodology selection was still necessary.	(Sangwan and Mittal 2015)
5	2016	Ifeyinwa Orji and Sun Wei	The authors discussed how manufacturing companies must implement green production techniques and expand product complexity while keeping costs down. Identifying the project costs for green manufacturing.	(Orji and Wei 2016)
6	2016	Suresh Prasad, Dinesh Khanduja, and Surrender K. Sharma	The authors investigated how GM approaches for waste reduction and labor efficiency measures could be applied in Indian foundries to increase production. The foundry industry would benefit from using lean and green techniques, according to findings.	(Prasad et al. 2016)
7	2016	Dinesh Seth, Rakesh Shrivastava, and Sanjeev Shrivastava	In order to develop and investigate a framework for green manufacturing based on identified vital success criteria and performance metrics, the study placed the Indian cement industry in perspective. With regard to green manufacturing, it offers a comprehensive framework that links top management, human resource management, corporate culture, best practices, project management, and supply chain management to performance indicators.	(Seth et al. 2016)

SN	Year	Author(s)	Findings	References
8	2016	Abhijeet K. Digalwar, Ashok R. Tagalpallewar, and Vivek K. Sunnapwar	The study discovered 12 distinct output counts for green manufacturing, each of which was constructed with its 66 constituent parts. Top management engagement, information management, employee training, green product and process design, staff empowerment, production planning, quality, cost, performance criteria for the consumer environment, customer reactions, and business growth were among these factors.	(Digalwar et al. 2016)
9	2017	Sarbjeet Kaushal, Dheeraj Gupta, and Hiralal Bhowmick	In the current work, martensitic stainless steel (SS-420) was coated with a wear-resistant composite cladding made of Ni-based+10% SiC using a newly discovered technology called microwave hybrid heating (MHH). In the case of the clad samples, wear is seen to result from particle displacement, tribofilm smearing, and craters caused by carbide pullout from the matrix.	(Kaushal et al. 2017)
10	2018	Sarbjeet Kaushal, Dheeraj Gupta, and Hiralal Bhowmick	The current study focuses on creating four functionally graded clads (FGC) of Ni-WC-based composite material on an AISI 304 substrate using microwave heating. The development of inter-metallics such NiW4, NiSi, and $Cr_{23}C_6$ was discovered by XRD analysis. The top FGC layer showed the highest microhardness value, which was 880 ± 30 HV.	(Kaushal et al. 2018a)
11	2018	Sarbjeet Kaushal, Dheeraj Gupta, and Hiralal Bhowmick	Through the use of a microwave hybrid heating process, Ni-based + 20% Cr_3C_2 composite clads on SS-304 austenitic stainless steel were created in the current work. The produced clads' average microhardness value was discovered to be 450 55 HV. Three times greater wear resistance is displayed by the microwave-processed clad than the SS-304 substrate.	(Kaushal et al. 2018b)

(continued)

SN	Year	Author(s)	Findings	References
12	2018	Sarbjeet Kaushal, Dheeraj Gupta & Hiralal Bhowmick	In the current study, austenitic stainless steel (SS-304) substrates were heated using a microwave hybrid heating method to produce functionally graded clads (FGC) of Ni-Cr_3C_2-based composite powders with variable percentages of Cr_3C_2 (0–30% by weight). A picture of the FGC taken using scanning electron microscopy (SEM) reveals that the Cr_3C_2 particles are evenly distributed throughout the Ni matrix. In the various layers of the FGC, Ni_3C, Ni_3Si, Ni_3Cr_2, and Cr_3C_2 phases were found.	(Kaushal et al. 2018c)
13	2019	Jeffrey A Bennett, Zachary S Campbell, and Milad Abolhasani	The authors of this work discuss cutting-edge flow chemistry techniques that can be used for highly valuable chemical synthesis that is also efficient in screening chemical reaction spaces. Less material and energy are required in a process thanks to the use of continuous production techniques.	(Bennett et al. 2019)
14	2019	Shuai Mao, Bing Wang, Yang Tang, and Feng Qian	Knowledge graphs, Bayesian networks, and deep learning are all parts of artificial intelligence, and authors have studied a number of enticing technologies in these areas. These tactics provide answers to important problems in green manufacturing.	(Mao et al. 2019)
15	2019	Guo Li, Ming K. Lim, and Zhaohua Wang	Investigated the link between performance and green production in the Chinese apparel sector. Stakeholders are unquestionably tied to green manufacturing practices, according to study findings.	(Li at al. 2019)
16	2019	Wei Dong Leong, Hon Loong Lam, Wendy Pei Qin Ng, Chun Hsion Lim, Chee Pin Tan, and Sivalinga Govinda Ponnambalam	Investigates the evolution and contribution of lean manufacturing and green manufacturing. Operations efficiency and the environment both benefit from lean and green practices.	(Leong et al. 2019)

SN	Year	Author(s)	Findings	References
17	2019	Sarbjeet Kaushal, Dilkaran Singh, Dheeraj Gupta, and Vivek Jain	The MMC clad of Ni + 10% WC8Co + Cr_3C_2-based material on SS-316 L substrate is effectively developed in the current work employing the reasonably priced microwave hybrid heating (MHH) technique. The results of the clad region's phase analysis investigation showed that different hard phases of $Co_3W_3C_4$, Cr_7Ni_3, NiC, NiW, W_2C, Fe_6W_6C, Fe_7C_3, and $FeNi_3$ formed after microwave heating.	(Kaushal et al. 2019a)
18	2019	Sarbjeet Kaushal, Dheeraj Gupta and Hiralal Bhowmick	The current study focuses on the creation of Ni/SiC-based functionally graded clads (FGC) employing microwave irradiation with different percentages of reinforced SiC from 0% to 30% (wt%) in the Ni-based powder. A maximum microhardness value of order 1020 ± 30 HV was found in the FGC top layer.	(Kaushal et al. 2019b)
19	2020	Ebenezer Afum, Osei-Ahenkan Victoria Yaa, Agyabeng-Mensah Yaw Amponsah, Owusu, Joseph Kusi, Lawrence Yaw Ankomah, and Joseph	Studied the illustrated relationship between green manufacturing practices (GMPs), sustainable performance, and green supply chain integration (GSCI) using data from a developing country. The findings demonstrate that implementing green manufacturing techniques has a significant influence on both sustainable performance and the incorporation of a green supply chain.	(Afum et al. 2020)
20	2020	B. Logesh and M. Balaji	Investigated the advantages of using lean manufacturing with green manufacturing. It is obvious that enterprises need lean manufacturing to increase resource utilization.	(Logesh and Balaji 2020)
21	2020	Guomin Lin and Botao Hao	Examined the current state of green manufacturing, its definition, organizational setup, associated technology, and potential future development trends. The authors conclude that the most urgent concern is environment safety after looking at a variety of factors.	(Lin and Hao 2020)

(continued)

SN	Year	Author(s)	Findings	References
22	2021	Sarbjeet Kaushal, Sourabh Bohra, Dheeraj Gupta, and Vivek Jain	The processing of microwave castings made of Cu-Mo-based materials was done in the authors' study utilizing microwave heating at 2.45 GHz and 900 W. The casting's electrical resistance increased as a result of the larger proportion of Mo particles inside the Cu-based matrix.	(Kaushal et al. 2021a)
23	2021	Sarbjeet Kaushal, Dheeraj Gupta, and Hiralal Bhowmick	In this study, WC_8Co compositions ranging from 0 to 30 wt% were treated employing a cost- and energy-efficient microwave hybrid heating approach on functionally graded clad layers of Ni-WC_8Co-based materials. Due to its superior mechanical qualities, it was found that functionally graded clad that had undergone microwave processing displayed wear resistance that was 95% and 29% higher than that of SS-304 substrate and Ni + 30%WC_8Co-based single-layer clad, respectively.	(Kaushal et al. 2021b)
24	2022	Jatinder Pal, Dheeraj Gupta, and Tejinder Singh	In the present study, iron-based alloy metal matrix composites (SS316) are successfully cast using a unique processing method called microwave hybrid heating (MHH). Without reinforcement, the microwave-cast SS316 is found to have a maximum relative density of 98%. The composite samples with microwave casting have a grain size value of 23 microns (μm).	(Pal et al. 2022)
25	2022	Anshika Prakash, Meenal Arora, Amit Mittal, Shivani Kampani, and Saurav Dixit	This study aims to pinpoint nine such barriers that pose problems for the adoption of green manufacturing. The research revealed a significant increase in publications on green manufacturing.	(Prakash et al. 2022)
26	2022	Mohammed A. Al-Hakimi, Abdullah Kaid Al-Swidi, Hamid Mahmood Gelaidan, and Abdulalem Mohammed	This study examined how GMP contributes to CSP through GI and evaluated the moderating and moderation-mediating influences of GOC using cross-sectional survey data from the selected SMEs in Saudi Arabia. Perhaps the most fascinating discovery was the empirical proof of GOC's moderating influence on GMP and GI along.	(Hakimi et al. 2022)

SN	Year	Author(s)	Findings	References
27	2022	Sarbjeet Kaushal, Satnam Singh, and Dheeraj Gupta	In the work the author researched, a method of processing the high-strength hybrid Ni-based composite covered in SS 316L steel surface and containing 15% (WC-8Co) and 5% Mo is done using microwave radiation. In comparison to the steel substrate, the clad that had undergone microwave processing showed a noticeable improvement in wear resistance.	(Kaushal et al. 2022)
28	2023	Yuzhen Chen, Xiaojun Ma, Xuejiao Ma, Meichen Shen, and Jingquan Chen	This study aims to establish the existence of a green premium in relation to a company's green transformation. On the basis of the results, two conclusions are provided, First, green transformation can improve stock returns in the short term by creating green premiums. Second, green transformation can boost stock returns by improving investors.	(Chen et al. 2023)
29	2023	Theresa Marie Sohns, Banu Aysolmaz, Lukas Figge, and Anant Joshi	The findings demonstrate the level of SMEs' commercial viability. The authors also highlight significant benefits, requirements, and challenges that might promote or restrict the usage of Green BPM practices. In light of the findings, authors make recommendations for how SMEs may accelerate the transition to higher levels of business sustainability.	(Sohns et al. 2023)
30	2023	Mukesh Kumar, Rahul Jain, Vikrant Sharma, and M.L. Meena	This study makes an effort to list, evaluate, rank, and model the key enablers. On the basis of a thorough review of the literature and professional opinion, the study identified 11 enablers.	(Kumar et al. 2023)
31	2023	Abid Haleem, Mohd Javaid, Ravi Pratap Singh, Rajiv Suman, and Mohd Asim Qadri	The article gives a general review of green manufacturing and emphasizes how important it is to preserve the environment. At the moment, it may be observed in many areas of our lives, such as recycling, energy-efficient technology, renewable energy sources, green building, and eco-friendly transportation.	(Haleem et al. 2023)

Explanation of this table is given in the following:

Fore and Mbohwa (2015): The authors observed the cement industry as it underwent a continuous process in order to incorporate green manufacturing into the manufacturing process or operations. The study investigated how the cement industry impacts the environment in a low-income, developing country with a specific focus on the sector. In a case study approach, a method emphasizing cleaner production was used, with the major focus being on problems like particle pollution and hazardous gases.

Dubey and Ali (2015): The authors examined how Indian businesses adopted green manufacturing techniques and how this affected the effectiveness of the wider supply chain. In order to reduce any significant impact of non-responsive bias on statistical analysis, wave analyses were also carried out to confirm non-response bias. Two waves of data collection were conducted. Utilizing this data, varimax rotation factor analysis was tested. Regression analysis was also influenced by the findings of the factor analysis.

Vamsi et al. (2015): The author examined the increasingly common practice known as "green product development." The main aim of developing green products is to lessen the collapse of industrial expansion. When developing green products, they observe a very haphazard usage of resources. The attempt has identified 80 different traits and 11 pillars using data from a related study to provide a framework for "green product development."

Sangwan and Mittal (2015): The author conducted research on the frameworks and themes associated to green development, including their origins, meanings, scope, and commonalities. He found that while many researchers had combined all of the models and approaches, there was still a need for standardization in the technique selection. Researchers needed to standardize their study terms, including the explicit use of various lifecycle engineering approaches, consistency regarding the implementation of end-of-life methods, incorporation of the entire supply chain, and alignment of environmental sustainability initiatives with corporate strategies; it was discovered during the literature review.

Orji and Wei (2016): The authors talked about how manufacturing firms must adopt green production practices and increase product complexity while maintaining an affordable price. Finding the project expenses related to green manufacturing, however, is a significant issue for engineering managers. Therefore, the engineer needs a system for anticipating and tracking the costs of green manufacturing.

Prasad et al. (2016): The author looked into how green manufacturing methods of labor efficiency measures and waste reduction could be employed in Indian foundries to boost productivity. Findings unambiguously show that adopting lean and green strategies is a good idea for the foundry business. This paper gives confirmation that lean and green practices are widely applicable in the casting industry and examines how these practices can be used in underdeveloped countries.

Seth et al. (2016): The study put the Indian cement industry in context in order to create and explore a framework for green manufacturing based on identified critical success criteria and performance measures. The authors' chosen approach for gathering data was a survey. Regression, proper statistical quality control methods, and factor analysis of the acknowledged key success variables are all utilized. It provides a complete framework for green manufacturing that connects top management, human resource management, corporate culture, best practices, project management, and supply chain management to performance indicators.

Digalwar et al. (2016): They examine the 128-item survey that was made with the help of literature research, interviews, and suggestions from nine different manufacturing companies, notably those that specialize in green manufacturing. The research identified a total of 12 green manufacturing output counts, which were established with their 66 components. These components included top management participation, information management, staff preparation, green product and process design, staff empowerment, production planning, quality, expense, performance standards for the consumer environment, customer responses, and business growth. It looked at 107 real replies from different Indian industries; its shortcoming is the inadequate sampling.

Kaushal et al. (2017): In the current work, martensitic stainless steel (SS-420) was coated with a wear-resistant composite cladding made of Ni-based+10% SiC using a newly discovered technology called microwave hybrid heating (MHH). For the development of clads in the current experiment, a household microwave oven with a frequency of 2.45 GHz and 900 W power was utilized. The clad's average Vicker microhardness was 652 ± 90 HV. Pin-on-disk sliding against an EN-31 (HRC-62) has been used to evaluate the tribological behavior of cladding. The clad surface demonstrated strong sliding wear resistance. In the case of the clad samples, wear is seen to result from particle displacement, tribofilm smearing, and craters caused by carbide pullout from the matrix.

Kaushal et al. (2018a): The current study focuses on creating four functionally graded clads (FGC) of Ni-WC-based composite material on an AISI 304 substrate using microwave heating. At 2.45 GHz, experimental tests were carried out within a domestic-style microwave applicator. Microstructural characterization results showed that the FGC was generated with a thickness of about 1.8 mm and was free of all types of interfacial cracks and apparent porosity. It was discovered that the WC particles in the nickel matrix were scattered at random. The development of inter-metallics such NiW4, NiSi, and $Cr_{23}C_6$ was discovered by XRD analysis. The top FGC layer showed the highest microhardness value, which was 880 ± 30 HV.

Kaushal et al. (2018b): Through the use of a microwave hybrid heating process, Ni-based + 20% Cr3C2 composite clads on SS-304 austenitic stainless steel were created in the current work. At 2.45 GHz and 900 W, experimental tests were carried out inside a home microwave appliance. Through the use of XRD, SEM/EDS, and Vicker's microhardness tests, the generated microwave composite clads were given a characterization. The XRD investigation verified the presence of FeNi3, NiSi, Cr3Ni2, and chromium carbide (Cr3C2), which helps to increase the microhardness of the composite clad. The produced clads' average microhardness value was discovered to be 450 55 HV. Three times greater wear resistance is displayed by the microwave-processed clad than the SS-304 substrate.

Kaushal et al. (2018c): In the current study, austenitic stainless steel (SS-304) substrates were heated using a microwave hybrid heating method to produce functionally graded clads (FGC) of Ni-Cr_3C_2-based composite powders with variable percentages of Cr_3C_2 (0–30% by weight). The experimental trials were carried out using a household microwave oven operating at 2.45 GHz with a variable power output of 180–900 W. The exposure period was adjusted to the compositional gradient. A picture of the FGC taken using scanning electron microscopy (SEM) reveals that the Cr_3C_2 particles are evenly distributed throughout the Ni matrix. In the various layers of the FGC, Ni_3C, Ni_3Si, Ni_3Cr_2, and Cr_3C_2 phases were found.

Bennett et al. (2019): Recently, the emphasis has switched to strengthening the sustainability and safety of chemical processes for both basic and applied research as well as the manufacture of pharmaceuticals and specialty chemicals. Flow chemistry techniques have attracted a lot of interest in an effort to create and improve chemical processes and satisfy the growing demand for chemical sustainability. The authors of this paper discuss innovative flow chemistry technologies that can be utilized for high-value chemical synthesis that is both environmentally friendly and effective at screening chemical reaction spaces. Continuous manufacturing techniques are being utilized more and more to reduce the amount of material and energy needed in a process. They incorporate real-time analysis, control, and increased process safety. Flow screening approaches can quickly search a multidimensional reaction space to improve process design, performance, and efficiency beyond continuous production. Additionally, by adopting time- and resource-effective (green) flow screening platforms, the next generation of process development technologies, including predictive reaction models and process scale-up techniques, can be created.

S. Mao et al. (2019): The significance, current state, and significant issues relating to green manufacturing in the process industry were thoroughly explored by the authors. Authors have examined a number of alluring technologies in knowledge graphs, Bayesian networks, and deep learning which are all components of AI. These strategies offer solutions for significant issues in green manufacturing. Specific technological problems for process security were covered after thorough investigation and discussions. These difficulties include the gathering of knowledge and the analysis of sparse error data, the precise fusing of heterogeneous data, early warning, and assisted decision-making. Theoretical solutions to these problems were put forth, and related successes were addressed.

Li et al. (2019): The authors investigated the relationship between green manufacturing and performance in China's fashion industry with a focus on assisting in raising environmental awareness and promoting green manufacturing methods. According to study findings, stakeholders are undoubtedly connected to green manufacturing practices. Enterprises might be inspired to pursue green manufacturing and environmental safety by the government and external stakeholders.

Leong et al. (2019): They examine the development and contribution of green and lean manufacturing. The primary contribution of this article is its concentration on the use and impact of the lean and green approaches on the manufacturing sector. The advantages and parallels between lean and green techniques were also investigated, as well as the numerous difficulties and misconceptions that industries confront. Lean and green techniques have a favorable impact on the performance of operations and the environment. The study also shows that industrialists and the government lack assistance and certainty in implementing green and lean approaches.

Kaushal et al. (2019a): The MMC clad of Ni + 10% WC8Co + Cr_3C_2-based material on SS-316 L substrate is effectively developed in the current work employing the reasonably priced microwave hybrid heating (MHH) technique. In the Ni-based matrix, the generated composite clads displayed a sophisticated microstructure with the presence of randomly distributed reinforcement particles. The results of the clad region's phase analysis investigation showed that different hard phases of $Co_3W_3C_4$, Cr_7Ni_3, NiC, NiW, W_2C, Fe_6W_6C, Fe_7C_3, and $FeNi_3$ formed after microwave heating. The hard phases that were present in the clad region helped to increase the microhardness. The clad region's mean microhardness was measured to be 550±40 HV. Furthermore, under various tribological circumstances, composite clad demonstrated superior wear resistance than SS-316 L substrate.

Kaushal et al. (2019b): The current study focuses on the creation of Ni/SiC-based functionally graded clads (FGC) employing microwave irradiation with different percentages of reinforced SiC from 0% to 30% (wt%) in the Ni-based powder. A household microwave applicator with a 2.45 GHz frequency range and a variable power output of 180–900 W was used to process the FGC. Back scattered electron (BSE) microscopy, energy dispersive spectroscopy (EDS), x-ray diffraction (XRD) spectroscopy, and Vicker's microhardness measurement were used to characterize the so-processed FGCs. The microstructural findings demonstrated the creation of FGC layers with various compositions. During the XRD examination, several phases like Ni_2Si, Ni_3C, NiSi, SiC, and $Cr_{23}C_6$ were seen to develop. A maximum microhardness value of order 1020 ± 30 HV was found in the FGC top layer. Furthermore, a three-point bend test was used to gauge the flexural strength of FGC. With a deformation index of 2.7×10^{-4} mmN^{-1}, the average flexural strength of FGC was measured to be 771.2 ± 5 MPa.

Afum et al. (2020): They used data from a poor nation to research the illustrative relationship between green manufacturing practices (GMPs), sustainable performance, and green supply chain integration (GSCI). The results show that adopting green manufacturing practices has a notable impact on both sustainable performance and the integration of a green supply chain. Additionally, the study shows that the inclusion of a green supply chain helps to bridge the gap among GMP and SP.

Logesh and Balaji (2020): They studied the benefits of accomplishing green manufacturing and lean manufacturing. Lean manufacturing is clearly necessary to improve resource utilization in industries. Reduce, reuse, and recycle were found to produce effective outcomes for green manufacturing.

Lin and Hao (2020): They analyzed the state of green manufacturing at the moment, its definition, organizational structure, related technologies, and future development patterns. After examining the numerous aspects, authors come to the conclusion that environment safety is the most pressing issue. The globalization of green manufacturing will be achieved through the addition, parallel, intellect, industrialization, and socialization of green manufacturing, according to authors. They also recommend that a perfect green manufacturing system be built.

Kaushal et al. (2021a): The processing of microwave castings made of Cu-Mo-based materials was done in the authors' study utilizing microwave heating at 2.45 GHz and 900 W. Inside the microwave applicator cavity, three separate casting sets were created, each with a different composition, such as Cu + 10% Mo, Cu + 30% Mo, and Cu + 50% Mo by volume percentage. It has been determined how microwave metal interaction works. A microstructure investigation of developed castings showed that Mo particles were uniformly dispersed in a Cu-based metal matrix, forming tiny, equiaxed grains. The phase analysis of composite castings that had undergone microwave processing revealed the existence of $Cu_{64}O$, $Cu_6Mo_5O_{18}$, and MoO_2 components phases. The Cu + 50% Mo-based casting had a maximum Vicker's microhardness of order 120.8 9 HV, which was around 2.7 times that of Cu-based castings that had undergone microwave processing. The casting's electrical resistance increased as a result of the larger proportion of Mo particles inside the Cu-based matrix.

Kaushal et al. (2021b): In this study, WC_8Co compositions ranging from 0 to 30 wt% were treated employing a cost- and energy-efficient microwave hybrid heating approach on functionally graded clad layers of Ni-WC_8Co-based materials. For comparing various attributes, single-layer clads with compositions corresponding to various functionally graded clad layers were also generated using microwave heating. Investigated were the effects of various sliding velocities and sliding distances on the wear characteristics of functionally graded clads produced by microwave processing. Due to its superior mechanical qualities, it was found that functionally graded clad that had undergone microwave processing displayed wear resistance that was 95% and 29% higher than that of SS-304 substrate and Ni + 30%WC_8Co-based single-layer clad, respectively.

The authors studied the most efficient variables that control how green manufacturing is carried out. The authors examine a number of variables including green manufacturing execution challenges, the role of the government in promoting green manufacturing, eco-knowledge, society effects, economic incentive, and innovation.

Pal et al. (2022): In the present study, iron-based alloy metal matrix composites (SS316) are successfully cast using a unique processing method called microwave hybrid heating (MHH). The iron-base alloy (SS316) castings are created in a household microwave oven with a power of 900 W and a frequency of 2.45 GHz and reinforced with 10 weight percent EWAC1004 (Ni-based) and 3 weight percent WC-12Co. Investigations are conducted in the prepared casts' mechanical and metallurgical characteristics. The obtained microhardness of cast samples is two times that of SS316 obtained commercially. The surface of the carbide phase exhibits a microhardness of 648.16 HV0.3. The prepared cast's porosity is discovered to be less than 2%, which is around three times less than SS316 that was purchased commercially. Without reinforcement, the microwave-cast SS316 is found to have a maximum relative density of 98%. The composite samples with microwave casting have a grain size value of 23 microns (μm).

Prakash et al. (2022): Based on the literature and the opinions of practitioners, this research report tries to identify nine such hurdles which affect the implementation of green manufacturing. The findings showed that there has been a considerable rise in publications on green manufacturing. This study also goes into detail about potential roadblocks to the adoption of GMP in the industrial industry.

Hakimi et al. (2022): This study used cross-sectional survey data from the chosen SMEs in Saudi Arabia to examine how GMP contributes to CSP through GI and assess the moderating and moderation-mediating impacts of GOC. According to the empirical data, the implementation of GMP significantly and favorably influences GI, which in turn greatly affects CSP. The empirical evidence of the moderating effect of GOC on the GMP and GI along was perhaps the most exciting finding.

Kaushal et al. (2022): In this work, the authors researched a method of processing the high-strength hybrid Ni-based composite covered in SS 316L steel surface and containing 15% (WC-8Co) and 5% Mo is done using microwave radiation. The premixed composite powder was applied to the steel surface and exposed to domestic microwave radiation at 900 W power and 2.45 GHz fixed frequency. Within 12–15 minutes of microwave irradiation, the hybrid clad was successfully generated. The clad that had undergone microwave processing displayed a flexural strength of 852 ± 6 MPa and a deformation index of 35 × 105 mm/N. The sliding wear investigation also showed that the production of oxide tribo layers was favored at a sliding velocity of 1.0 m/s. Abrasion, adhesion, surface pullout, and surface deformation were the several mechanisms of wear that were discovered through the research on worn-out surfaces. In comparison to the steel substrate, the clad that had undergone microwave processing showed a noticeable improvement in wear resistance.

Chen et al. (2023): The purpose of this study is to determine whether a green premium exists in relation to a firm's green transformation. Based on the findings, the following recommendations can be made. First, by introducing green premiums, green transformation can help enhance stock returns in the short run. Green action transformation can have a long-lasting effect, whereas green strategy transformation has less of an impact. Second, by enhancing investors, green transformation can increase stock returns.

Sohns et al. (2023): Due to their significant role in contributing to global pollution, small and medium-sized businesses (SMEs) have a crucial role to play in ensuring global sustainability. For SMEs, determining their level of company sustainability is difficult and specifies steps to lessen the impact on the environment. For this goal, sustainability assessment models have been created; however, they lack a process focus. Instead of focusing simply on end products and services, green business process management (green BPM) strives to lessen the detrimental environmental effect of organizations through business processes. Based on six Green BPM factors—green mentality, strategy, governance, modeling, monitoring, and optimization—the results show the degree of business sustainability of SMEs. Authors also point out pertinent advantages, demands, and obstacles that may encourage or inhibit the use of Green BPM practices. We offer suggestions on how SMEs might hasten the transition to higher levels of commercial sustainability in light of the findings.

Kumar et al. (2023): In order to encourage the adoption of green manufacturing practices in Ethiopian manufacturing businesses, this innovative study attempts to identify, examine, rank, and model the main enablers. The study identified 11 enablers based on a thorough evaluation of the literature and expert opinion. A novel combined FAHP-ISM-MICMAC analysis is used to rank the revealed enablers and analyze their interrelationships. The most important enablers are "customer awareness, pressure, and support" and "energy and resource crisis."

Haleem et al. (2023): This paper provides an overview of green manufacturing and highlights its significance in creating a sustainable environment. Critical green manufacturing applications for a sustainable environment are identified and explored, as well as many strategic tools and specialist methodologies for green manufacturing for an environmentally friendly future. Currently, it may be seen in many aspects of our lives, including recycling, energy-efficient technology, renewable energy sources, green construction, and environmentally friendly transportation. This offers fresh insight into integrated processes for product design and sustainable production.

9.10 VARIOUS ELEMENTS THAT INFLUENCE THE EXECUTION OF GREEN MANUFACTURING

a. **Implementation issues with green manufacturing**

    i. Costly products: Developing green products requires significant investment and creativity.
    ii. Ignorance: The general public continues to lack knowledge regarding the significance and advantages of becoming green.
    iii. Investment: Creating innovative technology is necessary for green products.

b. **The role of law in promoting green manufacturing:** The role of law in promoting green manufacturing is to create a legal framework that encourages environmentally friendly and sustainable business practices in the manufacturing sector. The law can establish requirements for lowering emissions of greenhouse gases, utilizing renewable energy sources, cutting waste, and conserving natural resources. Manufacturers must abide by environmental rules and regulations in many nations to guarantee that their production methods are environmentally friendly and sustainable. Penalties and fines for breaking these rules can be used to enforce them. Governments may also provide subsidies to businesses that use green manufacturing techniques.

c. **Organizational culture:** The implementation of green manufacturing practices depends critically on the organizational culture. Organizations can achieve their environmental sustainability goals through fostering a culture that emphasizes sustainability, employee involvement, continual improvement, education and training, and awards and recognition.

   i. **Leadership:** The atmosphere and course of an organization's culture are greatly influenced by its leadership. Leaders who place a high priority on environmental responsibility and sustainability can motivate their teams to use green manufacturing techniques.
   ii. **Employee engagement:** Companies that respect employee involvement and engagement are more likely to be successful in implementing green manufacturing techniques. Sustainable business practices are more likely to be supported and adopted by staff members who feel empowered to participate in environmental initiatives.
   iii. **Continuous enhancement:** Organizations can identify and address areas where they can improve their environmental performance with the aid of a culture that values continuous improvement. Companies can develop a culture of continuous improvement that drives the implementation of green manufacturing by encouraging staff to find possibilities for sustainability enhancements.
   iv. **Training and education:** Implementing green industrial practices requires education and training. Employees can acquire the skills and knowledge necessary to successfully adopt sustainable practices by working in an organization where continuous education and training are prioritized.
   v. **Recognition and remuneration:** Employees may be inspired to accept sustainable manufacturing practices by a culture that values and promotes sustainability initiatives. It might help to emphasize the significance of these efforts and foster a culture of sustainability inside the company to acknowledge and award personnel for their efforts to environmental sustainability.

d. **Eco-Knowledge:** Ecological knowledge is essential for putting green manufacturing into practice. A procedure known as "green manufacturing" aims to minimize the negative effects of manufacturing operations on the environment. Organizations must have a thorough awareness of sustainability-related environmental challenges, laws, and advances in technology in order to successfully execute green manufacturing. Ecological knowledge can be useful in this situation. Organizations can benefit from eco-knowledge by the following ways:

i. Determine how they can cut down on waste and energy use.
ii. Select ecologically friendly items and sustainable materials.
iii. Create novel, ecologically friendly, and effective manufacturing techniques.
iv. Develop cutting-edge, ecologically friendly goods and services.
v. Keep abreast on environmental laws and standards.
vi. Inform employees and other interested parties on environmental problems and sustainable practices.

e. **Corporate environment:** The company climate can have a big impact on how well green manufacturing is implemented. Green manufacturing practices are more likely to be effectively implemented in an organization that is dedicated to sustainability and has a culture, policies, and practices that prioritize environmental responsibility.

i. **Policies and practices:** An organization's policies and practices can influence its sustainability strategy. Green manufacturing practices are more likely to be used by a company, for instance, if it has policies in place to decrease waste, conserve energy, and lower its carbon footprint.
ii. **Technology investment:** Investing in new machinery and technology is frequently necessary for green production. There is a greater probability that an organization that is dedicated to sustainability will invest in these technologies and use them in its production procedures.
iii. **Employee education:** The skills and expertise of employees are crucial to the success of green manufacturing. Green manufacturing can be successfully implemented through training and educating personnel on sustainable practices.
iv. **Supply chain management:** Companies may guarantee that their vendors are putting sustainable practices into place by working with them. In order to achieve this, suppliers may be encouraged to utilize eco-friendly products, and logistics and transportation may be optimized.

f. **Society's impact:** The adoption of green manufacturing techniques is significantly influenced by society. Society may play a significant role in supporting the adoption of green manufacturing practices by creating consumer interest in sustainable products, raising awareness, and working with many stakeholders.

i. **Demand from consumers:** Demand for green products is largely driven by consumers. Consumers are increasingly inclined to look for products created utilizing sustainable manufacturing practices as they grow more environmentally concerned. In order to satisfy the demands of customers, this may motivate businesses to embrace green production techniques.
ii. **Government regulations:** By putting laws and incentives into place, governments may assist in promoting green industrial practices. Governments may, for instance, provide tax incentives or grants to businesses that implement sustainable practices, or they may levy fines on businesses that don't adhere to ecological standards.

iii. **Public awareness:** Through a variety of avenues, including the media, NGOs, and social networking sites, society may educate people on the significance of sustainable manufacturing practices. As stakeholders and customers learn more about how production affects the environment, pressure on businesses to switch to sustainable practices may increase.
iv. **Collaboration and partnerships:** Collaboration and partnerships among many stakeholders, including governmental organizations, business associations, and environmental groups, can help to advance green manufacturing techniques. For instance, industry organizations can collaborate with businesses to create sustainability standards and guidelines, and environmental advocacy groups can offer knowledge and support to compel businesses to adopt sustainable practices.

g. Innovations: Green innovation in goods satisfies customer requirements for protecting the environment, aids businesses in expanding into new markets, hinders competitors from replicating products, and keeps products competitive. The following variables affect the adoption of green innovations:

i. Green techniques and modern technologies.
ii. Participation by stakeholders.
iii. Dependability and specifications quality.
iv. Taking responsibility and performing responsibly.
v. Systems for evaluating and guidance.

Generally speaking, the first and second components are more significant; emphasis varies among groups of stakeholders.

9.11 CONCLUSION

This review study revealed the following parameters that were the most effective or responsible in the context of green manufacturing:

1. Problems implementing green manufacturing
2. The function of law in advancing green manufacturing
3. Organizational culture
4. Eco-knowledge
5. Corporate environment
6. Society's impact
7. Financial reward
8. Innovation

REFERENCES

Afum, E., Osei-Ahenkan, V.Y., Agyabeng-Mensah, Y., Amponsah Owusu, J., Kusi, L.Y., & Ankomah, J. (2020). Green manufacturing practices and sustainable performance among Ghanaian manufacturing SMEs: the explanatory link of green supply chain integration. *Management of Environmental Quality: An International Journal*, 31(6) 1457–1475.

Ahn, D. G. (2016). Direct metal addictive manufacturing processes and their sustainable applications for green technology. *International Journal of Precision Engineering and Manufacturing Green Technology*, 3(4), 381–395.

Al-Hakimi, M. A., Al-Swidi, A. K., Gelaidan, H. M., & Mohammed, A. (2022). The influence of green manufacturing practices on the corporate sustainable performance of SMEs under the effect of green organizational culture: A moderated mediation analysis. *Journal of Cleaner Production*, 376, 13434.

Bennett, J. A., Campbell, Z. S., & Milad, A. (2019). Role of continuous flow processes in green manufacturing of pharmaceuticals and specialty chemicals. *Current Opinion in Chemical Engineering*, 26, 9–19.

Bhanot, N., Rao, P. V., & Deshmukh S.G. (2017). An integrated approach for analysing the enablers and barriers of sustainable manufacturing, *Journal of Cleaner Production*, 142(4), 4412–4439.

Chen, Y., Ma, X., Ma, X., Shen, M., & Chen. J. (2023). Does green transformation trigger green premiums? Evidence from Chinese listed manufacturing firms. *Journal of Cleaner Production*, 407, 136858.

Digalwar, A. K., Nidhi, M., Tagalpallewar, A. R., & Sunnapwar, V. K. (2017). Roadmap for the implementation of green manufacturing practices in Indian manufacturing industries: an ISM approach. *Benchmarking an International Journal*, 24(5), 1386–1399.

Digalwar, A. K., Tagalpallewar, A. R., & Sunnapwar, V. K. (2013). Green manufacturing performance measures: an empirical investigation from Indian manufacturing industries. *Measuring Business Excellence*, 17(4), 59–75.

Dubey, R., & Ali, S. S. (2015). Exploring antecedents of extended supply chain performance measures: an insight from Indian green manufacturing practices, *Benchmarking an International Journal*, 22(5), 752–772.

Fore, S., & Mbohawa, C. (2015). Greening manufacturing practices in a continuous process industry: case study of a cement manufacturing company. *Journal of Engineering Design and Technology*, 13(1), 94–122.

Haleem, A., Javaid, M., Singh, R. P., Suman, R., & Qadri, M. S. (2023). A pervasive study on Green Manufacturing towards attaining sustainability. *Green Technologies and Sustainability*, 1(2), 100018.

IEA (2021), *Global Energy Review 2021*, IEA, Paris. www.iea.org/reports/global-energy-review-2021.

Kaushal, S., Bohra, S., & Gupta, D. (2021a). On processing and characterization of Cu–Mo-based castings through microwave heating. *Inter Metalcast,* 15, 530–537.

Kaushal, S., Gupta, D., & Bhowmick, H. (2017a). On microstructure and wear behavior of microwave processed composite clad. *ASME Journal of Tribology*, 139(6), 061602.

Kaushal, S., Gupta, D., & Bhowmick, H. (2017b). Investigation of dry sliding wear behavior of Ni–SiC microwave cladding. *ASME Journal of Tribology*, 139(4), 041603.

Kaushal, S., Gupta, D., & Bhowmick, H. (2018a). An approach for functionally graded cladding of composite material on austenitic stainless steel substrate through microwave heating. *Journal of Composite Materials*, 52(3), 301–312.

Kaushal, S., Gupta, D., & Bhowmick, H. (2018b). On processing of Ni-WC based functionally graded composite clads through microwave heating. *Materials and Manufacturing Processes*, 33(8), 822–828.

Kaushal, S., Gupta, D., & Bhowmick, H. (2018c). On surface modification of austenitic stainless steel using microwave processed Ni/Cr_3C_2 composite cladding, *Surface Engineering*, 34(11), 809–817.

Kaushal, S., Gupta, D., & Bhowmick, H. (2018d). On processing of Ni-Cr_3C_2 based functionally graded clads through microwave heating. *Materials Research Express,* 5, 066405.

Kaushal, S., Gupta, D., & Bhowmick, H. (2019a). On processing and flexural behaviour of functionally graded clads developed through microwave irradiation. *Materials Research Express,* 6, 076405.

Kaushal, S., Gupta, D., & Bhowmick, H. (2021b). Wear behavior of microwave-processed Ni-WC_8Co-based functionally graded materials. Proceedings of the Institution of Mechanical Engineers. *Part L: Journal of Materials: Design and Applications*, 235(5), 1036–1045.

Kaushal, S., Singh, S., & Gupta, D. (2022). Processing strategy for high strength Ni-based hybrid composite clad on SS 316L steel through microwave heating. Proceedings of the Institution of Mechanical Engineers, *Part B: Journal of Engineering Manufacture,* 236(3), 190–203.

Kaushal, S., Singh, D., Gupta, D., & Jain, V. (2019b). Wear resistance improvement of Austenitic 316 L steel by microwave-processed composite clads. *ASME Journal of Tribology*, 141(4), 041605.

Kumar, M., Jain, R., Sharma, V., & Meena, M. L. (2023). Modelling of green manufacturing enablers using integrated MCDM approach for Ethiopian manufacturing industry. *Materials Today: Proceedings.* https://doi.org/10.1016/j.matpr.2023.02.236

Lin, G., & Hao, B. (2020). Research on green green manufacturing technology. *Journal of Physics: Conference Series*, 1601(4), 042046.

Logesh, B., & Balaji, M. (2020). *International Journal of Precision Engineering and Manufacturing-Green Technology,* 40684-020-00216-4.

Logesh, B., & Balaji, M. (2021). Experimental investigations to deploy green manufacturing through reduction of waste using lean tools in electrical components manufacturing company. *International Journal of Precision Engineering and Manufacturing-Green Technology*, 8, 365–374.

Mao, S., Wang, B., Tang, Y., & Qian, F. (2019). Opportunities and challenges of artificial intelligence for green manufacturing in the process industry. *Engineering*, S2095809919300074.

Ministry of New and Renewable energy (2022), Available at https://pib.gov.in/FeaturesDeatils.aspx?NoteId=151141&ModuleId%20=%202.

Mohanty, R.P., Seth, D., & Mukadam, S. (2007). Quality dimensions of e-commerce and their implications. *Total Quality Management & Business Excellence*, 18(3), 219–247.

Orji, I., & Wei, S. (2016). A detailed calculation model for costing of green manufacturing. *Industrial Management & Data Systems*, 116(1), 65–86.

Pal, J., Gupta, D., & Singh, T. (2022). Processing and characterization of SS316 based metal matrix composite casting through microwave hybrid heating. Proceedings of the Institution of Mechanical Engineers, *Part C: Journal of Mechanical Engineering Science.* 236. 095440622211044. 10.1177/09544062221104443.

Paul, I. D., Bhole, G. P., & Chaudhari, J. R. (2014). A review on green manufacturing: it's important, methodology and its application. *Procedia Materials Science,* 6, 1644–1649.

Prakash, A., Arora, M., Mittal, A., Kampani, S., & Dixit, S. (2022). Green manufacturing: related literature over the past decade, *Materials Today: Proceedings,* 69(2), 468–472.

Prasad, S., Kinduja, D., & Sharma, S. K. (2016). An empirical study on applicability of lean and green practices in the foundry industry. *Journal of Manufacturing Technology Management,* 27, 2–29.

Pulicherla, K. K., Adapa, V., Ghosh, M., & Ingle, P. (2022). Current efforts on sustainable green growth in the manufacturing sector to complement "make in India" for making "self-reliant India." *Environmental Research*, 206, 112263.

Rehman, A. A., Minhaj, R., Shrivastava, R., & Shrivastava R. L. (2013). Validating green manufacturing (GM) framework for sustainable development in an Indian steel industry. *Universal Journal of Mechanical Engineering*, 1(2), 49–61.

Sangwan, K. S. (2006). Performance value analysis for justification of green manufacturing systems. *Journal of Advanced Manufacturing Systems*, 5(1), 59–73.

Seth, D., & Gupta, V. (2005). Application of value stream mapping for lean operations and cycle time reduction: an Indian case study. *Production Planning & Control*, 16(1), 44–59.

Seth, D., & Panigrahi, A. (2015). Application and evaluation of packaging postponement strategy to boost supply chain responsiveness: a case study. *Production Planning & Control*, 26(13), 1069–1089.

Seth, D., Shrivastava, R., & Shrivastava, S. (2016). An empirical investigation of critical success factors and performance measures for green manufacturing in cement industry. *Journal of Manufacturing Technology Management*, 27(8), 1076–1101.

Singh, J., Singh, C. D., & Deepak D. (2022). Recent trends in industrial and production engineering, 978-981-16-3135.

Singh, S. K., & Mittal, V. K. (2015). A bibliometric analysis of green manufacturing and similar frameworks. *Management of Environmental Quality: An International Journal*, 26(4), 566–587.

Sohns, T. M., Aysolmaz, B., Figge, L., & Joshi, A. (2023). Green business process management for business sustainability: A case study of manufacturing small and medium-sized enterprises (SMEs) from Germany. *Journal of Cleaner Production*, 401, 136667.

Thanki, S., Govindan, K., & Thakkar, J. (2016). An investigation on lean-green implementation practices in Indian SMEs using analytical hierarchy process (AHP) approach. *Journal of Cleaner Production*, 135, 284–298.

Tianyi, G. (2014). Research on Green Manufacturing Model and Key Technologies of Automobile Enterprises [D]. *Jilin University*.

Toke L. K., & Kalpande S. D. (2019). Critical success factors of green manufacturing for achieving sustainability in Indian context, *International Journal of Sustainable Engineering*, 12(6), 415–422.

Vamsi, N. J. K., Sharma, A., & Shashikantha, K. (2015). Development of a framework for green product development. *Benchmarking: An International Journal,* 22(3), 426–445.

Vellinga, P., & Herb, N. (1999). International Human Dimensions Programme on Global Environmental Change, Bonn, Germany, softcover. *Resources, Conservation and Recycling*, 31(1), 105–106.

Xianguang, L. (2014). Carbon emission analysis of dry gear processing for green manufacturing [J]. *China Mechanical Engineering*, 25(16), 2184–2190.

Yinbao, N. (2014). Research on tool evaluation scheme for green manufacturing [J]. *Combined Machine Tool and Automatic Processing Technology*, 8, 1–4.

Zheneng, W. (2010). Development direction of modern manufacturing – green manufacturing [J]. *Equipment Manufacturing Technology,* 03, 35–38.

Index

N

O

P

R

S

T

U

W

Z

www.ingramcontent.com/pod-product-compliance
Lightning Source LLC
LaVergne TN
LVHW010605110826
845149LV00003B/777
9781032582412